STUDENT SOLUTION

Introduction to Statistical Quality Control

Seventh Edition

Douglas C. Montgomery
Arizona State University

Prepared by

Cheryl Jennings
Bank of America

WILEY

COVER ILLUSTRATION: **Norm Christiansen**

ISBN   978-1-118-57359-4

Printed in the United States of America

10 9 8 7 6 5 4 3 2 1

Printed and bound by Hamilton Printing Company

Acknowledgments and Additional Information

The following software was used to prepare the exercise solutions for *Introduction to Statistical Quality Control, 7th Edition*, by Douglas C. Montgomery:

- Minitab Statistical Software, Release 16 for Windows. Minitab Inc., State College, PA. MINITAB® and all other trademarks and logos for the Company's products and services are the exclusive property of Minitab Inc. All other marks referenced remain the property of their respective owners. See minitab.com for more information. Portions of information contained in this publication/book are printed with permission of Minitab Inc. All such material remains the exclusive property and copyright of Minitab Inc. All rights reserved.
- Microsoft® Excel 2010. Microsoft Corporation, Redmond, WA (www.microsoft.com).

The following Minitab and Excel data files and solutions are available from the companion website for the textbook associated with this supplement. To reach the student companion website, go to www.wiley.com/college/montgomery, select the main text, then click on the "student companion site" link.

Chapter	Student Data Files
1	No files
2	No files
3	Chap 03-S.xlsx
4	Chap 04-S.xlsx
5	Chap 05-S.xlsx
6	Chap 06-S.xlsx
7	Chap 07-S.xlsx
8	Chap 08-S.xlsx
9	Chap 09-S.xlsx
10	Chap 10-S.xlsx
11	Chap 11-S.xlsx
12	Chap 12-S.xlsx
13	Chap 13-S.xlsx, Chap 13-S.mpj
14	Chap 14-S.xlsx, Chap 14-S.mpj
15	Chap 15-S.xlsx
16	Chap 16-S.xlsx

Suggestion or corrections should be directed to the textbook author:

Prof. Douglas C. Montgomery
School of Computing, Informatics and Decision Systems Engineering
Arizona State University
Tempe, AZ 85287
Email: doug.montomgery@asu.edu

Chapter 3

Modeling Process Quality

LEARNING OBJECTIVES

After completing this chapter you should be able to:

1. Construct and interpret visual data displays, including the stem-and-leaf plot, the histogram, and the box plot
2. Compute and interpret the sample mean, the sample variance, the sample standard deviation, and the sample range
3. Explain the concepts of a random variable and a probability distribution
4. Understand and interpret the mean, variance, and standard deviation of a probability distribution
5. Determine probabilities from probability distributions
6. Understand the assumptions for each of the discrete probability distributions presented
7. Understand the assumptions for each of the continuous probability distributions presented
8. Select an appropriate probability distribution
9. Use probability plots
10. Use approximations for some hypergeometric and binomial distributions

IMPORTANT TERMS AND CONCEPTS

Approximations to probability distributions	Percentile
Binomial distribution	Poisson distribution
Box plot	Population
Central limit theorem	Probability distribution
Continuous distribution	Probability plotting
Descriptive statistics	Quartile
Discrete distribution	Random variable
Exponential distribution	Run chart
Gamma distribution	Sample
Geometric distribution	Sample average
Histogram	Sample standard deviation
Hypergeometric probability distribution	Sample variance

Interquartile range   Standard deviation
Lognormal distribution   Standard normal distribution
Mean of a distribution   Statistics
Median   Stem-and-leaf display
Negative binomial distribution   Time series plot
Normal distribution   Uniform distribution
Normal probability plot   Variance of a distribution
Pascal distribution   Weibull distribution

EXERCISES

3.1.

The content of liquid detergent bottles is being analyzed. Twelve bottles, randomly selected from the process, are measured, and the results are as follows (in fluid ounces): 16.05, 16.03, 16.02, 16.04, 16.05, 16.01, 16.02, 16.02, 16.03, 16.01, 16.00, 16.07.

(a) Calculate the sample average.

$$\bar{x} = \sum_{i=1}^{n} x_i \bigg/ n = \left(16.05 + 16.03 + \cdots + 16.07\right)\big/12 = 16.029 \text{ oz}$$

(b) Calculate the sample standard deviation.

$$s = \sqrt{\frac{\sum_{i=1}^{n} x_i^2 - \left(\sum_{i=1}^{n} x_i\right)^2 \bigg/ n}{n-1}} = \sqrt{\frac{\left(16.05^2 + \cdots + 16.07^2\right) - \left(16.05 + \cdots + 16.07\right)^2 \big/12}{12-1}} = 0.0202 \text{ oz}$$

MTB > Stat > Basic Statistics > Display Descriptive Statistics
Descriptive Statistics: Ex3-1

Variable	N	N*	Mean	SE Mean	StDev	Minimum	Q1	Median	Q3
Ex3-1	12	0	16.029	0.00583	0.0202	16.000	16.012	16.025	16.047

Variable	Maximum
Ex3-1	16.070

3.5.

The nine measurements that follow are furnace temperatures recorded on successive batches in a semiconductor manufacturing process (units are °F): 953, 955, 948, 951, 957, 949, 954, 950, 959

(a) Calculate the sample average.

$$\bar{x} = \sum_{i=1}^{n} x_i \bigg/ n = \left(953 + 955 + \cdots + 959\right)/9 = 952.9 \ °F$$

(b) Calculate the sample standard deviation.

$$s = \sqrt{\frac{\sum_{i=1}^{n} x_i^2 - \left(\sum_{i=1}^{n} x_i\right)^2 \big/ n}{n-1}} = \sqrt{\frac{\left(953^2 + \cdots + 959^2\right) - \left(953 + \cdots + 959\right)^2 \big/ 9}{9-1}} = 3.7 \ °F$$

MTB > Stat > Basic Statistics > Display Descriptive Statistics

Descriptive Statistics: Ex3-5

Variable	N	N*	Mean	SE Mean	StDev	Minimum	Q1	Median	Q3
Ex3-5	9	0	952.89	1.24	3.72	948.00	949.50	953.00	956.00

Variable	Maximum
Ex3-5	959.00

3.7.

Yield strengths of circular tubes with end caps are measured. The first yields (in kN) are as follows: 96, 102, 104, 108, 126, 128, 150, 156

(a) Calculate the sample average.

$$\bar{x} = \sum_{i=1}^{n} x_i \Big/ n = (96 + 102 + \cdots + 156)/8 = 121.25 \text{ kN}$$

(b) Calculate the sample standard deviation.

$$s = \sqrt{\frac{\sum\limits_{i=1}^{n} x_i^2 - \left(\sum\limits_{i=1}^{n} x_i\right)^2 \Big/ n}{n-1}} = \sqrt{\frac{(96^2 + \cdots + 156^2) - (96 + \cdots + 156)^2 \big/ 8}{8-1}} = 22.63 \text{ kN}$$

```
MTB > Stat > Basic Statistics > Display Descriptive Statistics
```
Descriptive Statistics: Ex3-7

```
Variable   N   N*    Mean   SE Mean   StDev   Minimum      Q1  Median      Q3
Ex3-7      8   0   121.25      8.00   22.63     96.00   102.50  117.00  144.50

Variable   Maximum
Ex3-7       156.00
```

3.15.

Consider the viscosity data in Exercise 3.10. Construct a normal probability plot, a lognormal probability plot, and a Weibull probability plot for these data. Based on the plots, which distribution seems to be the best model for the viscosity data?

MTB > Graph > Probability Plot > Simple and select Normal Distribution

Probability Plot of Ex3-10 / Table 3E.3

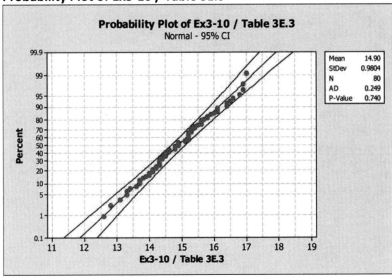

3.15. continued

MTB > Graph > Probability Plot > Simple and select Lognormal Distribution
Probability Plot of Ex3-10 / Table 3E.3

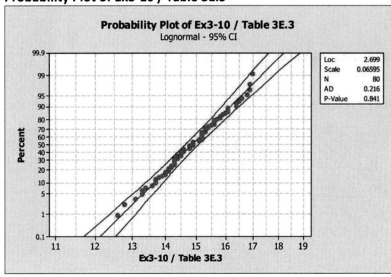

3.15. continued

MTB > Graph > Probability Plot > Simple and select Weibull Distribution
Probability Plot of Ex3-10 / Table 3E.3

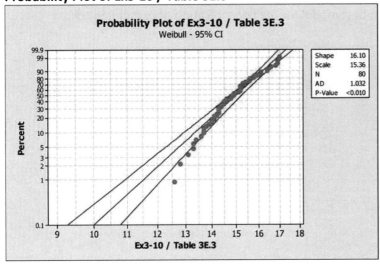

Both the normal and lognormal distributions appear to be reasonable models for the data; the plot points tend to fall along a straight line, with no bends or curves. However, the plot points on the Weibull probability plot are not straight—particularly in the tails—indicating it is not a reasonable model.

3.17.

An important quality characteristic of water is the concentration of suspended solid material (in ppm). Table 3E.5 contains 40 measurements on suspended solids for a certain lake. Construct a normal probability plot, a lognormal probability plot, and a Weibull probability plot for these data. Based on the plots, which distribution seems to be the best model for the concentration of suspended solids?

▪ **TABLE 3E.5**
Concentration of Suspended Solids (ppm)

0.78	9.59	2.26	8.13	3.16
4.33	11.70	0.22	125.93	1.30
0.15	0.20	0.29	13.72	0.96
0.29	2.93	3.65	3.47	1.73
14.21	1.79	0.54	14.81	0.68
0.09	5.81	5.17	21.01	0.41
4.75	2.82	1.30	4.57	74.74
0.78	1.94	3.52	20.10	4.98

MTB > Graph > Probability Plot > Simple and select Normal Distribution
Probability Plot of Ex3-17 / Table 3E.5

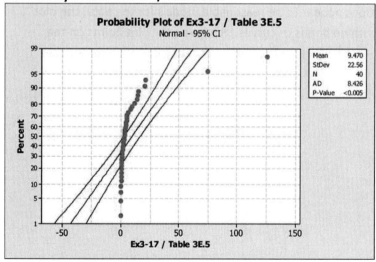

3.17. continued

MTB > Graph > Probability Plot > Simple and select Lognormal Distribution
Probability Plot of Ex3-17 / Table 3E.5

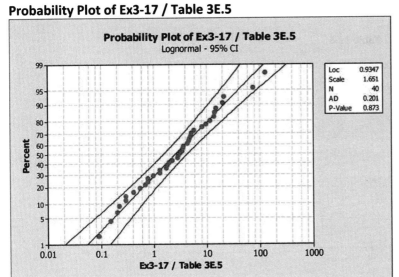

3.17. continued

MTB > Graph > Probability Plot > Simple and select Weibull Distribution
Probability Plot of Ex3-17 / Table 3E.5

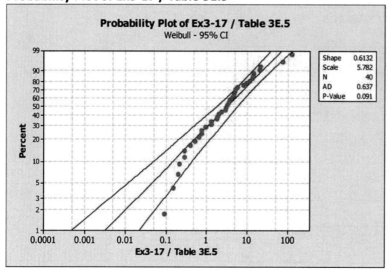

The lognormal distribution appears to be a reasonable model for the concentration data. Plotted points on the normal and Weibull probability plots tend to fall off a straight line.

3.23.

Consider the chemical process yield data in Exercise 3.9. Calculate the sample average and standard deviation.

$$\bar{x} = \sum_{i=1}^{n} x_i \Big/ n = \frac{94.1 + 93.2 + \cdots + 84.5}{90} = 89.476$$

$$S = \sqrt{\frac{\sum_{i=1}^{n} x_i^2 - \left(\sum_{i=1}^{n} x_i\right)^2 \Big/ n}{n-1}} = \sqrt{\frac{(94.1^2 + \cdots + 84.5^2) - \dfrac{(94.1 + \cdots + 84.5)^2}{90}}{90-1}} = 4.158$$

```
MTB > Stat > Basic Statistics > Display Descriptive Statistics
Descriptive Statistics: Ex3-9 / Table 3E.2

Variable              N    N*     Mean    SE Mean   StDev   Minimum      Q1    Median
Ex3-9 / Table 3E.2    90    0   89.476     0.438   4.158    82.600   86.100   89.250

Variable                   Q3   Maximum
Ex3-9 / Table 3E.2    93.125    98.000
```

3.27.

Suppose that two fair dice are tossed and the random variable observed—say, x—is the sum of the two up faces. Describe the sample space of this experiment, and determine the probability distribution of x.

x: {the sum of two up faces}
sample space: {2, 3, 4, 5, 6, 7, 8, 9, 10, 11, 12}

Calculate the probability of rolling a 2: obtained by rolling a 1 on each die:

$$Pr\{x=2\}=Pr\{1,1\}=\frac{1}{6}\times\frac{1}{6}=\frac{1}{36}$$

Calculate the probability of rolling a 3: 1 and 2 or 2 and 1:

$$Pr\{x=3\}=Pr\{1,2\}+Pr\{2,1\}=\left(\frac{1}{6}\times\frac{1}{6}\right)+\left(\frac{1}{6}\times\frac{1}{6}\right)=\frac{2}{36}$$

Calculate the probability of rolling a 4:

$$Pr\{x=4\}=Pr\{1,3\}+Pr\{2,2\}+Pr\{3,1\}=\left(\frac{1}{6}\times\frac{1}{6}\right)+\left(\frac{1}{6}\times\frac{1}{6}\right)+\left(\frac{1}{6}\times\frac{1}{6}\right)=\frac{3}{36}$$

. . .

$$p(x)=\begin{cases} 1/36; x=2 \\ 2/36; x=3 \\ 3/36; x=4 \\ 4/36; x=5 \\ 5/36; x=6 \\ 6/36; x=7 \\ 5/36; x=8 \\ 4/36; x=9 \\ 3/36; x=10 \\ 2/36; x=11 \\ 1/36; x=12 \\ 0; \text{otherwise} \end{cases}$$

3.29.

A mechatronic assembly is subjected to a final functional test. Suppose that defects occur at random in these assemblies, and that defects occur according to a Poisson distribution with parameter $\lambda = 0.02$.

(a) What is the probability that an assembly will have exactly one defect?

This is a Poisson distribution with parameter $\lambda = 0.02$, $x \sim POI(0.02)$.

$$\Pr\{x = 1\} = p(1) = \frac{e^{-0.02}(0.02)^1}{1!} = 0.0196$$

(b) What is the probability that an assembly will have one or more defects?

$$\Pr\{x \geq 1\} = 1 - \Pr\{x = 0\} = 1 - p(0) = 1 - \frac{e^{-0.02}(0.02)^0}{0!} = 1 - 0.9802 = 0.0198$$

(c) Suppose that you improve the process so that the occurrence rate of defects is cut in half to $\lambda = 0.01$. What effect does this have on the probability that an assembly will have one or more defects?
This is a Poisson distribution with parameter $\lambda = 0.01$, $x \sim POI(0.01)$.

$$\Pr\{x \geq 1\} = 1 - \Pr\{x = 0\} = 1 - p(0) = 1 - \frac{e^{-0.01}(0.01)^0}{0!} = 1 - 0.9900 = 0.0100$$

Cutting the rate at which defects occur reduces the probability of one or more defects by approximately one-half, from 0.0198 to 0.0100.

3.31.

The random variable x takes on the values 1, 2, or 3 with probabilities $(1 + 3k)/3$, $(1 + 2k)/3$, and $(0.5 + 5k)/3$, respectively.

$$p(x) = \begin{cases} (1+3k)/3; & x=1 \quad (1+2k)/3; \quad x=2 \\ (0.5+5k)/3; & x=3 \quad 0; \quad \text{otherwise} \end{cases}$$

(a) Find the appropriate value of k.

To solve for k, use $F(x) = \sum_{i=1}^{\infty} p(x_i) = 1$

$$\frac{(1+3k)+(1+2k)+(0.5+5k)}{3} = 1$$

$$10k = 0.5$$

$$k = 0.05$$

(b) Find the mean and variance of x.

$$\mu = \sum_{i=1}^{3} x_i p(x_i) = 1 \times \left[\frac{1+3(0.05)}{3} \right] + 2 \times \left[\frac{1+2(0.05)}{3} \right] + 3 \times \left[\frac{0.5+5(0.05)}{3} \right] = 1.867$$

$$\sigma^2 = \sum_{i=1}^{3} x_i^2 p(x_i) - \mu^2 = 1^2(0.383) + 2^2(0.367) + 3^2(0.250) - 1.867^2 = 0.615$$

(c) Find the cumulative distribution function.

$$F(x) = \begin{cases} \dfrac{1.15}{3} = 0.383; & x=1 \\[2mm] \dfrac{1.15+1.1}{3} = 0.750; & x=2 \\[2mm] \dfrac{1.15+1.1+0.75}{3} = 1.000; & x=3 \end{cases}$$

3.33.

A manufacturer of electronic calculators offers a one-year warranty. If the calculator fails for any reason during this period, it is replaced. The time to failure is well modeled by the following probability distribution:

$$f(x) = 0.125e^{-0.125x} \quad x > 0$$

(a) What percentage of the calculators will fail within the warranty period?

This is an exponential distribution with parameter $\lambda = 0.125$: $\Pr\{x \leq 1\} = F(1) = 1 - e^{-0.125(1)} = 0.118$.
Approximately 11.8% will fail during the first year.

(b) The manufacturing cost of a calculator is $50, and the profit per sale is $25. What is the effect of warranty replacement on profit?

Mfg. cost = $50/calculator and Sale profit = $25/calculator
Net profit = $[-50(1 + 0.118) + 75]/calculator = $19.10/calculator.
The effect of warranty replacements is to decrease profit by $5.90/calculator.

3.35.

A production process operates with 1% nonconforming output. Every hour a sample of 25 units of product is taken, and the number of nonconforming units counted. If one or more nonconforming units are found, the process is stopped and the quality control technician must search for the cause of nonconforming production. Evaluate the performance of this decision rule.

This is a binomial distribution with parameter $p = 0.01$ and $n = 25$. The process is stopped if $x \geq 1$.

$$\Pr\{x \geq 1\} = 1 - \Pr\{x < 1\} = 1 - \Pr\{x = 0\} = 1 - \binom{25}{0}(0.01)^0(1-0.01)^{25} = 1 - 0.78 = 0.22$$

This decision rule means that 22% of the samples will have one or more nonconforming units, and the process will be stopped to look for a cause. This is a somewhat difficult operating situation.

3.35. continued

This exercise may also be solved using Excel (Excel Function BINOMDIST(x, n, p, TRUE)) or Minitab.

MTB > Calc > Probability Distribution > Binomial
Cumulative Distribution Function

```
Binomial with n = 25 and p = 0.01

x   P( X <= x )
0      0.777821
```

3.37.

A random sample of 50 units is drawn from a production process every half hour. The fraction of nonconforming product manufactured is 0.02. What is the probability that $\hat{p} \leq 0.04$ if the fraction nonconforming really is 0.02?

This is a binomial distribution with parameter $p = 0.02$ and $n = 50$.

$$\Pr\{\hat{p} \leq 0.04\} = \Pr\{x \leq 2\} = \sum_{x=0}^{4} \binom{50}{x} (0.02)^x (1-0.02)^{(50-x)}$$

$$= \binom{50}{0}(0.02)^0(1-0.02)^{50} + \binom{50}{1}(0.02)^1(1-0.02)^{49} + \cdots + \binom{50}{4}(0.02)^4(1-0.02)^{46} = 0.921$$

Therefore, the probability is 0.921 that the sample fraction nonconforming can be $\hat{p} \leq 0.04$ if the population fraction nonconforming really is $p = 0.02$.

3.43.

An electronic component for a medical X-ray unit is produced in lots of size $N = 25$. An acceptance testing procedure is used by the purchaser to protect against lots that contain too many nonconforming components. The procedure consists of selecting five components at random from the lot (without replacement) and testing them. If none of the components is nonconforming, the lot is accepted.

(a) If the lot contains two nonconforming components, what is the probability of lot acceptance?

This is a hypergeometric distribution with $N = 25$ and $n = 5$, without replacement.
Given $D = 2$ and $x = 0$:

$$\Pr\{Acceptance\} = p(0) = \frac{\binom{2}{0}\binom{25-2}{5-0}}{\binom{25}{5}} = \frac{(1)(33,649)}{(53,130)} = 0.633$$

This exercise may also be solved using Excel (Excel Function HYPGEOMDIST(x, n, D, N)) or Minitab.

```
MTB > Calc > Probability Distribution > Hypergeometric
Cumulative Distribution Function

Hypergeometric with N = 25, M = 2, and n = 5

x    P( X <= x )
0       0.633333
```

(b) Calculate the desired probability in (a) using the binomial approximation. Is this approximation satisfactory? Why or why not?

For the binomial approximation to the hypergeometric, $p = D/N = 2/25 = 0.08$ and $n = 5$.

$$\Pr\{acceptance\} = p(0) = \binom{5}{0}(0.08)^0(1-0.08)^5 = 0.659$$

This approximation, though close to the exact solution for $x = 0$, violates the rule-of-thumb that $n/N = 5/25 = 0.20$ be less than the suggested 0.1. The binomial approximation is not satisfactory in this case.

3.43. continued

(c) Suppose the lot size was $N = 150$. Would the binomial approximation be satisfactory in this case?

For $N = 150$, $n/N = 5/150 = 0.033 \le 0.1$, so the binomial approximation would be a satisfactory approximation the hypergeometric in this case.

(d) Suppose that the purchaser will reject the lot with the decision rule of finding one or more nonconforming components in a sample of size n, and wants the lot to be rejected with probability at least 0.95 if the lot contains five or more nonconforming components. How large should the sample size n be?

Find n to satisfy $\Pr\{x \ge 1 \mid D \ge 5\} \ge 0.95$, or equivalently $\Pr\{x = 0 \mid D = 5\} < 0.05$.

$$p(0) = \frac{\binom{5}{0}\binom{25-5}{n-0}}{\binom{25}{n}} = \frac{\binom{5}{0}\binom{20}{n}}{\binom{25}{n}}$$

try $n = 10$

$$p(0) = \frac{\binom{5}{0}\binom{20}{10}}{\binom{25}{10}} = \frac{(1)(184,756)}{(3,268,760)} = 0.057$$

try $n = 11$

$$p(0) = \frac{\binom{5}{0}\binom{20}{11}}{\binom{25}{11}} = \frac{(1)(167,960)}{(4,457,400)} = 0.038$$

Let sample size $n = 11$.

3.45.

A textbook has 500 pages on which typographical errors could occur. Suppose that there are exactly 10 such errors randomly located on those pages. Find the probability that a random selection of 50 pages will contain no errors. Find the probability that 50 randomly selected pages will contain at least two errors.

This is a hypergeometric distribution with N = 500 pages, n = 50 pages, and D = 10 errors. Checking n/N = 50/500 = 0.1 $\leq$ 0.1, the binomial distribution can be used to approximate the hypergeometric, with p = D/N = 10/500 = 0.020.

$$\Pr\{x=0\}=p(0)=\binom{50}{0}(0.020)^0(1-0.020)^{50-0}=(1)(1)(0.364)=0.364$$

$$\Pr\{x\geq2\}=1-\Pr\{x\leq1\}=1-[\Pr\{x=0\}+\Pr\{x=1\}]=1-p(0)-p(1)$$

$$=1-0.364-\binom{50}{1}(0.020)^1(1-0.020)^{50-1}=1-0.364-0.372=0.264$$

3.47.

Glass bottles are formed by pouring molten glass into a mold. The molten glass is prepared in a furnace lined with firebrick. As the firebrick wears, small pieces of brick are mixed into the molten glass and finally appear as defects (called "stones") in the bottle. If we can assume that stones occur randomly at the rate of 0.00001 per bottle, what is the probability that a bottle selected at random will contain at least one such defect?

This is a Poisson distribution with λ = 0.00001 stones/bottle.

$$\Pr\{x\geq1\}=1-\Pr\{x=0\}=1-\frac{e^{-0.00001}(0.00001)^0}{0!}=1-0.99999=0.00001$$

3.49.

A production process operates in one of two states: the in-control state, in which most of the units produced conform to specifications, and an out-of-control state, in which most of the units produced are defective. The process will shift from the in-control to the out-of-control state at random. Every hour, a quality control technician checks the process, and if it is in the out-of-control state, the technician detects this with probability p. Assume that when the process shifts out of control it does so immediately following a check by the inspector, and once a shift has occurred, the process cannot automatically correct itself. If t denotes the number of periods the process remains out of control following a shift before detection, find the probability distribution of t. Find the mean number of periods the process will remain in the out-of-control state.

The distribution for t is based on the number of periods for which the process shifted from in control to out of control, 1, with probability $(p)^1$, and the periods from which the process remains in the out of control state without a shift, $t - 1$, with probability $(1 - p)^{t-1}$. There is only one permutation for this situation. Hence, the distribution is the combination of the two states as follows:

$Pr(t) = p(1-p)^{t-1}$; $t = 1,2,3,...$ The mean is calculated using (3.5b):

$$\mu = \sum_{i=1}^{\infty} x_i p(x_i) \Rightarrow \sum_{t=1}^{\infty} t \left[p(1-p)^{t-1} \right] = p \frac{d}{dq} \left[\sum_{t=1}^{\infty} q^t \right] = \frac{1}{p}$$

3.51.

The tensile strength of a metal part is normally distributed with mean 40 pounds and standard deviation 5 pounds. If 50,000 parts are produced, how many would you expect to fail to meet a minimum specification limit of 35-pounds tensile strength? How many would have a tensile strength in excess of 48 pounds?

$x \sim N(40, 5^2)$; $n = 50{,}000$

How many fail the minimum specification, LSL = 35 lb.? Using the standard normal distribution, Appendix II:

$$Pr\{x \le 35\} = Pr\left\{z \le \frac{35-40}{5}\right\} = Pr\{z \le -1\} = \Phi(-1) = 0.159$$

The number that fail the minimum specification are $(50{,}000) \times (0.159) = 7950$.

This exercise may also be solved using Excel (Excel Function NORMDIST(X, μ, σ, TRUE)) or Minitab.

```
MTB > Calc > Probability Distribution > Normal
Cumulative Distribution Function

Normal with mean = 40 and standard deviation = 5

  x   P( X <= x )
 35      0.158655
```

How many exceed 48 lb.?

$$Pr\{x > 48\} = 1 - Pr\{x \le 48\} = 1 - Pr\left\{z \le \frac{48-40}{5}\right\} = 1 - Pr\{z \le 1.6\} = 1 - \Phi(1.6) = 1 - 0.945 = 0.055$$

The number that exceed 48 lb are $(50{,}000) \times (0.055) = 2750$.

3.53.

Continuation of Exercise 3.52. Reconsider the power supply manufacturing process in Exercise 3.52. Suppose we wanted to improve the process. Can shifting the mean reduce the number of nonconforming units produced? How much would the process variability need to be reduced in order to have all but one out of 1000 units conform to the specifications?

The process, with mean 5 V, is currently centered between the specification limits (target = 5 V). Shifting the process mean in either direction would increase the number of nonconformities produced. Desire Pr{Conformance} = 1 / 1000 = 0.001. Assume that the process remains centered between the specification limits at 5 V. Need Pr$\{x \le$ LSL$\}$ = 0.001 / 2 = 0.0005.

$$\Phi(z) = 0.0005$$
$$z = \Phi^{-1}(0.0005) = -3.29$$
$$z = \frac{\text{LSL} - \mu}{\sigma}, \quad \text{so } \sigma = \frac{\text{LSL} - \mu}{z} = \frac{4.95 - 5}{-3.29} = 0.015$$

Process variance must be reduced to 0.015^2 to have at least 999 of 1000 conform to specification.

3.55.

The life of an automotive battery is normally distributed with mean 900 days and standard deviation 35 days. What fraction of these batteries would be expected to survive beyond 1000 days?

$$x \sim N(900, 35^2)$$
$$\text{Pr}\{x > 1000\} = 1 - \text{Pr}\{x \le 1000\}$$
$$= 1 - \text{Pr}\left\{x \le \frac{1000 - 900}{35}\right\}$$
$$= 1 - \Phi(2.8571)$$
$$= 1 - 0.9979$$
$$= 0.0021$$

3.57.

The specifications on an electronic component in a target-acquisition system are that its life must be between 5000 and 10,000 h. The life is normally distributed with mean 7500 h. The manufacturer realizes a price of $10 per unit produced; however, defective units must be replaced at a cost of $5 to the manufacturer. Two different manufacturing processes can be used, both of which have the same mean life. However, the standard deviation of life for process 1 is 1000 h, whereas for process 2 it is only 500 h. Production costs for process 2 are twice those for process 1. What value of production costs will determine the selection between processes 1 and 2?

$x_1 \sim N(7500, \sigma_1^2 = 1000^2)$; $x_2 \sim N(7500, \sigma_2^2 = 500^2)$; LSL = 5,000 h; USL = 10,000 h

sales = $10/unit, defect = $5/unit, profit = $10 × Pr{good} + $5 × Pr{bad} − c

Under_line{For Process 1}

proportion defective $= p_1 = 1 - \Pr\{\text{LSL} \le x_1 \le \text{USL}\} = 1 - \Pr\{x_1 \le \text{USL}\} + \Pr\{x_1 \le \text{LSL}\}$

$$= 1 - \Pr\left\{ z_1 \le \frac{10,000 - 7,500}{1,000} \right\} + \Pr\left\{ z_1 \le \frac{5,000 - 7,500}{1,000} \right\}$$

$$= 1 - \Phi(2.5) + \Phi(-2.5) = 1 - 0.9938 + 0.0062 = 0.0124$$

profit for process 1 = 10 (1 − 0.0124) + 5 (0.0124) − c_1 = 9.9380 − c_1

Under_line{For Process 2}

proportion defective $= p_2 = 1 - \Pr\{\text{LSL} \le x_2 \le \text{USL}\} = 1 - \Pr\{x_2 \le \text{USL}\} + \Pr\{x_2 \le \text{LSL}\}$

$$= 1 - \Pr\left\{ z_2 \le \frac{10,000 - 7,500}{500} \right\} + \Pr\left\{ z_2 \le \frac{5,000 - 7,500}{500} \right\}$$

$$= 1 - \Phi(5) + \Phi(-5) = 1 - 1.0000 + 0.0000 = 0.0000$$

profit for process 2 = 10 (1 − 0.0000) + 5 (0.0000) − c_2 = 10 − c_2

If $c_2 > c_1 + 0.0620$, then choose process 1

Chapter 4

Inferences About Process Quality

LEARNING OBJECTIVES

After completing this chapter you should be able to:

1. Explain the concept of random sampling
2. Explain the concept of a sampling distribution
3. Explain the general concept of estimating the parameters of a population or probability distribution
4. Know how to explain the precision with which a parameter is estimated
5. Construct and interpret confidence intervals on a single mean and on the difference in two means
6. Construct and interpret confidence intervals on a single variance or the ratio of two variances
7. Construct and interpret confidence intervals on a single proportion and on the difference in two proportions
8. Test hypotheses on a single mean and on the differences in two means
9. Test hypotheses on a single variance and on the ratio of two variances
10. Test hypotheses on a single proportion and on the difference in two proportions
11. Use the P-value approach for hypothesis testing
12. Understand how the analysis of variance (ANOVA) is used to test hypotheses about the equality of more than two means
13. Understand how to fit and interpret linear regression models

IMPORTANT TERMS AND CONCEPTS

Alternative hypothesis	Point estimator
Analysis of variance (ANOVA)	Poisson distribution
Binomial distribution	Pooled estimator
Checking assumptions for statistical inference procedures	Power of a statistical test
Chi-square distribution	Random sample
Confidence interval	Regression model
Confidence intervals on means, known variance(s)	Residual analysis

Confidence intervals on means, unknown
   variance(s)

Confidence intervals on proportions

Confidence intervals on the variance of a normal
   distribution

Confidence intervals on the variances of two
   normal distributions

Critical region for a test statistic

F-distribution

Hypothesis testing

Least squares estimator

Linear statistical model

Minimum variance estimator

Null hypothesis

P-value

P-value approach

Parameters of a distribution

Sampling distribution

Scaled residuals

Statistic

t-distribution

Test statistic

Tests of hypotheses on means, known variance(s)

Tests of hypotheses on means, unknown
   variance(s)

Tests of hypotheses on proportions

Tests of hypotheses on the variance of a normal
   distribution

Tests of hypotheses on the variances of two
   normal distributions

Type I error

Type II error

Unbiased estimator

EXERCISES

4.1.

Suppose that you are testing the following hypotheses where the variance is known:

$H_0 : \mu = 100$
$H_1 : \mu \neq 100$

Find the P-value for the following values of the test statistic.

(a) $Z_0 = 2.75$

From Appendix II, $\Phi(Z_0) = \Phi(2.75) = 0.99702$
For a two-sided test: $P = 2[1 - \Phi(|Z_0|)] = 2[1 - 0.99702] = 0.00596$

(b) $Z_0 = 1.86$

From Appendix II, $\Phi(Z_0) = \Phi(1.86) = 0.96856$
For a two-sided test: $P = 2[1 - \Phi(|Z_0|)] = 2[1 - 0.96856] = 0.06289$

(c) $Z_0 = -2.05$

From Appendix II, $\Phi(Z_0) = \Phi(|-2.05|) = \Phi(2.05) = 0.97982$
For a two-sided test: $P = 2[1 - \Phi(|Z_0|)] = 2[1 - 0.97982] = 0.04036$

(d) $Z_0 = -1.86$

Same answer as for (b)
From Appendix II, $\Phi(Z_0) = \Phi(|-1.86|) = \Phi(1.86) = 0.96856$
For a two-sided test: $P = 2[1 - \Phi(|Z_0|)] = 2[1 - 0.96856] = 0.06289$

4.3.

Suppose that you are testing the following hypotheses where the variance is known:

$H_0 : \mu = 100$

$H_1 : \mu < 100$

Find the P-value for the following values of the test statistic.

(a) $Z_0 = -2.35$

From Appendix II, $\Phi(Z_0) = \Phi(2.35) = 0.99061$
For a one-sided, lower-tail test and by symmetry of the normal distribution:
$P = \Phi(Z_0) = \Phi(-2.35) = 1 - \Phi(2.35) = 1 - 0.99061 = 0.00939$

(b) $Z_0 = -1.99$

From Appendix II, $\Phi(Z_0) = \Phi(1.99) = 0.97670$
For a one-sided, lower-tail test and by symmetry of the normal distribution:
$P = \Phi(Z_0) = \Phi(-1.99) = 1 - \Phi(1.99) = 1 - 0.97670 = 0.02330$

(c) $Z_0 = -2.18$

From Appendix II, $\Phi(Z_0) = \Phi(2.18) = 0.98537$
For a one-sided, lower-tail test and by symmetry of the normal distribution:
$P = \Phi(Z_0) = \Phi(-2.18) = 1 - \Phi(2.18) = 1 - 0.98537 = 0.01463$

(d) $Z_0 = -1.85$

From Appendix II, $\Phi(Z_0) = \Phi(1.85) = 0.96784$
For a one-sided, lower-tail test and by symmetry of the normal distribution:
$P = \Phi(Z_0) = \Phi(-1.85) = 1 - \Phi(1.85) = 1 - 0.96784 = 0.03216$

4.5.

Suppose that you are testing the following hypotheses where the variance is unknown:

$H_0 : \mu = 100$

$H_1 : \mu > 100$

The sample size is $n = 12$. Find bounds on the P-value for the following values of the test statistic.

(a) $t_0 = 2.55$

From Appendix IV and $v = n - 1 = 12 - 1 = 11$,

α	0.025		0.01
$T_{\alpha, 11}$	2.20	2.55	2.718

For a one-sided, upper-tail test: $0.01 < P < 0.025$

(b) $t_0 = 1.87$

From Appendix IV and $v = 11$,

α	0.05		0.025
$T_{\alpha, 11}$	1.796	1.87	2.20

For a one-sided, upper-tail test: $0.025 < P < 0.05$

(c) $t_0 = 2.05$

From Appendix IV and $v = 11$,

α	0.05		0.025
$T_{\alpha, 11}$	1.796	2.05	2.20

For a one-sided, upper-tail test: $0.025 < P < 0.05$

(d) $t_0 = 2.80$

From Appendix IV and $v = 11$,

α	0.01		0.005
$T_{\alpha, 11}$	2.718	2.80	3.106

For a one-sided, upper-tail test: $0.005 < P < 0.01$

4.7.

The inside diameters of bearings used in an aircraft landing gear assembly are known to have a standard deviation of $\sigma = 0.002$ cm. A random sample of 15 bearings has an average inside diameter of 8.2535 cm.

(a) Test the hypothesis that the mean inside bearing diameter is 8.25 cm. Use a two-sided alternative and $\alpha = 0.05$.

Since σ is known, use the standard normal distribution:

$x \sim N(\mu, \sigma)$, $n = 15$, $\bar{x} = 8.2535$ cm, $\sigma = 0.002$ cm, $\mu_0 = 8.25$, $\alpha = 0.05$

Test H_0: $\mu = 8.25$ vs. H_1: $\mu \neq 8.25$. Reject H_0 if $|Z_0| > Z_{\alpha/2}$.

$$Z_0 = \frac{\bar{x} - \mu_0}{\sigma/\sqrt{n}} = \frac{8.2535 - 8.25}{0.002/\sqrt{15}} = 6.78 \quad \text{(Equation 4.23)}$$

$Z_{\alpha/2} = Z_{0.05/2} = Z_{0.025} = 1.96$ (from Appendix II)

Reject H_0: $\mu = 8.25$, and conclude that the mean bearing ID is not equal to 8.25 cm.

(b) Find the P-value for this test.

From Appendix II, find the corresponding cumulative standard normal, $\Phi(Z_0)$ for $Z0 = 6.78$, and calculate:

P-value $= 2[1 - \Phi(Z_0)] = 2[1 - \Phi(6.78)] = 2[1 - 1.00000] = 0$

(c) Construct a 95% two-sided confidence interval on the mean bearing diameter.

Construct a 95% two-sided confidence interval on mean bearing diameter.

$$\bar{x} - Z_{\alpha/2}\left(\sigma/\sqrt{n}\right) \leq \mu \leq \bar{x} + Z_{\alpha/2}\left(\sigma/\sqrt{n}\right)$$

$$8.25 - 1.96\left(0.002/\sqrt{15}\right) \leq \mu \leq 8.25 + 1.96\left(0.002/\sqrt{15}\right) \quad \text{(Equation 4.29)}$$

$$8.249 \leq \mu \leq 8.251$$

```
MTB > Stat > Basic Statistics > 1-Sample Z
One-Sample Z

The assumed standard deviation = 0.002

 N     Mean   SE Mean        95% CI
15   8.23530  0.00052  (8.23429, 8.23631)
```

4.9.

The service life of a battery used in a cardiac pacemaker is assumed to be normally distributed. A random sample of ten batteries is subjected to an accelerated life test by running them continuously at an elevated temperature until failure, and the following lifetimes (in hours) are obtained: 25.5, 26.1, 26.8, 23.2, 24.2, 28.4, 25.0, 27.8, 27.3, and 25.7.

(a) The manufacturer wants to be certain that the mean battery life exceeds 25 h. What conclusions can be drawn from these data (use $\alpha = 0.05$).

Since σ is unknown, calculate the sample standard deviation S, and use the t-distribution, one-sided:

$x \sim N(\mu, \sigma)$, $n = 10$, $\bar{x} = 26.0$, $s = 1.62$, $\mu_0 = 25$, $\alpha = 0.05$

Test H_0: $\mu = 25$ vs. H_1: $\mu > 25$. Reject H_0 if $t_0 > t_\alpha$.

$$t_0 = (\bar{x} - \mu_0)/(s/\sqrt{n}) = (26.0 - 25)/(1.62/\sqrt{10}) = 1.952 \quad \text{(Equation 4.33)}$$

$t_{\alpha, n-1} = t_{0.05, 10-1} = 1.833$ (from Appendix IV)

Reject H_0: $\mu = 25$, and conclude that the mean life exceeds 25 h.

```
MTB > Stat > Basic Statistics > 1-Sample t
One-Sample T: Ex 4-9

Test of mu = 25 vs > 25

                                      95% Lower
Variable    N    Mean   StDev   SE Mean    Bound     T      P
Ex 4-9      10   26.000  1.625   0.514    25.058   1.95   0.042
```

(b) Construct a 90% two-sided confidence interval on mean life in the accelerated test.

Construct a 90% two-sided confidence interval on mean life in the accelerated test. $\alpha = 0.10$

$$\bar{x} - t_{\alpha/2, n-1}\, s/\sqrt{n} \le \mu \le \bar{x} + t_{\alpha/2, n-1}\, s/\sqrt{n}$$

$$26.0 - 1.833(1.62/\sqrt{10}) \le \mu \le 26.0 + 1.833(1.62/\sqrt{10}) \quad \text{(Equation 4.34)}$$

$$25.06 \le \mu \le 26.94$$

4.9.(b) continued

```
MTB > Stat > Basic Statistics > 1-Sample t
One-Sample T: Ex 4-9

Test of mu = 25 vs not = 25

Variable    N    Mean    StDev   SE Mean        90% CI          T      P
Ex 4-9     10   26.000   1.625    0.514   (25.058, 26.942)   1.95   0.083
```

(c) Construct a normal probability plot of the battery life data. What conclusions can you draw?

MTB > Graph > Probability Plot > Single and ensure Distribution = Normal
Probability Plot of Ex 4-9

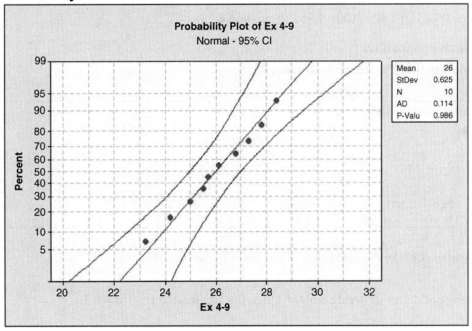

The plotted points fall approximately along a straight line, so the assumption that battery life is normally distributed is appropriate.

4.11.

A new process has been developed for applying photoresist to 125-mm silicon wafers used in manufacturing integrated circuits. Ten wafers were tested, and the following photoresist thickness measurements (in angstromx × 1000) were observed: 13.3987, 13.3957, 13.3902, 13.4015, 13.4001, 13.3918, 13.3965, 13.3925, 13.3946, and 13.4002.

(a) Test the hypothesis that mean thickness is 13.4 × 1000 Å. Use $\alpha = 0.05$ and assume a two-sided alternative.

Since σ is unknown, calculate the sample standard deviation S, and use the t-distribution:

$x \sim N(\mu, \sigma)$, $n = 10$, $\bar{x} = 13.39618 \times 1000$ Å, $s = 0.00391$, $\mu_0 = 13.4 \times 1000$ Å, $\alpha = 0.05$

Test H_0: $\mu = 13.4$ vs. H_1: $\mu \neq 13.4$. Reject H_0 if $|t_0| > t_{\alpha/2}$.

$$t_0 = \left(\bar{x} - \mu_0\right)\big/\left(s/\sqrt{n}\right) = \left(13.39618 - 13.4\right)\big/\left(0.00391/\sqrt{10}\right) = -3.089 \quad \text{(Equation 4.33)}$$

$t_{\alpha/2,\,n-1} = t_{0.025,\,9} = 2.262$ (from Appendix IV)

Reject H_0: $\mu = 13.4$, and conclude that the mean thickness differs from 13.4 × 1000 Å.

```
MTB > Stat > Basic Statistics > 1-Sample t
One-Sample T: Ex 4-11

Test of mu = 13.4 vs not = 13.4

Variable    N     Mean    StDev   SE Mean        95% CI             T      P
Ex 4-11    10   13.3962  0.0039   0.0012   (13.3934, 13.3990)   -3.09  0.013
```

(b) Find a 99% two-sided confidence interval on mean photoresist thickness. Assume that thickness is normally distributed.

$\alpha = 0.01$

$$\bar{x} - t_{\alpha/2,n-1}\left(s/\sqrt{n}\right) \leq \mu \leq \bar{x} + t_{\alpha/2,n-1}\left(s/\sqrt{n}\right)$$

$$13.39618 - 3.2498\left(0.00391/\sqrt{10}\right) \leq \mu \leq 13.39618 + 3.2498\left(0.00391/\sqrt{10}\right) \quad \text{(Equation 4.34)}$$

$$13.39216 \leq \mu \leq 13.40020$$

4.11.(b) continued

MTB > Stat > Basic Statistics > 1-Sample t
One-Sample T: Ex 4-11

```
Test of mu = 13.4 vs not = 13.4

Variable    N    Mean    StDev  SE Mean        99% CI          T      P
Ex 4-11    10  13.3962  0.0039  0.0012  (13.3922, 13.4002)  -3.09  0.013
```

(c) Does the normality assumption seem reasonable for these data?

MTB > Graph > Probability Plot > Single and ensure Distribution = Normal
Probability Plot of Ex 4-11

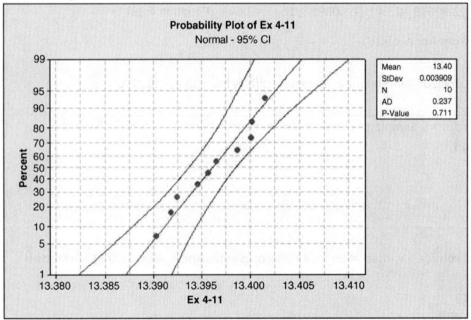

The plotted points form a reverse-"S" shape, instead of a straight line, so the assumption that battery life is normally distributed is not appropriate.

4.13.

Ferric chloride is used as a flux in some types of extraction metallurgy processes. This material is shipped in containers, and the container weight varies. It is important to obtain an accurate estimate of mean container weight. Suppose that from long experience a reliable value for the standard deviation of flux container weight is determined to be 4 lb. How large a sample would be required to construct a 95% two-sided confidence interval on the mean that has a total width of 1 lb?

$\sigma = 4$ lb, $\alpha = 0.05$, $Z_{\alpha/2} = Z_{0.025} = 1.9600$, total confidence interval width $= 1$ lb, find n

$$2\left[Z_{\alpha/2}\left(\sigma/\sqrt{n}\right)\right] = \text{total width}$$

$$2\left[1.9600\left(4/\sqrt{n}\right)\right] = 1$$

$$n = 246$$

4.15.

The output voltage of a power supply is assumed to be normally distributed. Sixteen observations taken at random on voltage are as follows: 10.35, 9.30, 10.00, 9.96, 11.65, 12.00, 11.25, 9.58, 11.54, 9.95, 10.28, 8.37, 10.44, 9.25, 9.38, and 10.85.

(a) Test the hypothesis that the mean voltage equals 12 V against a two-sided alternative using $\alpha = 0.05$.

$x \sim N(\mu, \sigma)$, $n = 16$, $\bar{x} = 10.259$ V, $s = 0.999$ V, $\mu_0 = 12$, $\alpha = 0.05$

Test $H_0: \mu = 12$ vs. $H_1: \mu \neq 12$. Reject H_0 if $|t_0| > t_{\alpha/2}$.

$$t_0 = \left(\bar{x} - \mu_0\right)/\left(s/\sqrt{n}\right) = \left(10.259 - 12\right)/\left(0.999/\sqrt{16}\right) = -6.971 \quad \text{(Equation 4.33)}$$

$t_{\alpha/2, n-1} = t_{0.025, 15} = 2.131$ (from Appendix IV)

Reject $H_0: \mu = 12$, and conclude that the mean output voltage differs from 12V.

```
MTB > Stat > Basic Statistics > 1-Sample t
One-Sample T: Ex 4-15

Test of mu = 12 vs not = 12

Variable    N     Mean   StDev   SE Mean      95% CI          T      P
Ex 4-15    16   10.259   0.999    0.250   (9.727, 10.792)   -6.97  0.000
```

14.15. continued

(b) Construct a 95% two-sided confidence interval on μ.

Construct a 95% two-sided confidence interval on μ.

$$\bar{x} - t_{\alpha/2,n-1}\left(S/\sqrt{n}\right) \le \mu \le \bar{x} + t_{\alpha/2,n-1}\left(S/\sqrt{n}\right)$$

$$10.259 - 2.131\left(0.999/\sqrt{16}\right) \le \mu \le 10.259 + 2.131\left(0.999/\sqrt{16}\right) \quad \text{(Equation 4.34)}$$

$$9.727 \le \mu \le 10.792$$

(c) Test the hypothesis that $\sigma^2 = 11$ using $\alpha = 0.05$.

Test H_0: $\sigma^2 = 1$ vs. H_1: $\sigma^2 \ne 1$. Reject H_0 if $\chi^2_0 > \chi^2_{\alpha/2,\,n-1}$ or $\chi^2_0 < \chi^2_{1-\alpha/2,\,n-1}$.

$$\chi^2_0 = (n-1)S^2/\sigma^2_0 = \frac{(16-1)0.999^2}{1} = 14.970 \quad \text{(Equation 4.38)}$$

$\chi^2_{\alpha/2,\,n-1} = \chi^2_{0.025,16-1} = 27.488$ (from Appendix III)

$\chi^2_{1-\alpha/2,\,n-1} = \chi^2_{0.975,16-1} = 6.262$ (from Appendix III)

Do not reject H_0: $\sigma^2 = 1$, and conclude that there is insufficient evidence that the variance differs from 1.

(d) Construct a 95% two-sided confidence interval on σ.

$$(n-1)S^2/\chi^2_{\alpha/2,n-1} \le \sigma^2 \le (n-1)S^2/\chi^2_{1-\alpha/2,n-1}$$

$$(16-1)0.999^2/27.488 \le \sigma^2 \le (16-1)0.999^2/6.262 \quad \text{(Equation 4.39)}$$

$$0.545 \le \sigma^2 \le 2.391$$

$$0.738 \le \sigma \le 1.546$$

14.15.(d) continued

Since the 95% confidence interval on σ contains the hypothesized value, $\sigma_0^2 = 1$, the null hypothesis, $H_0: \sigma^2 = 1$, cannot be rejected. Using MINITAB:

MTB > Stat > Basic Statistics > Graphical Summary
Summary for Ex 4-15

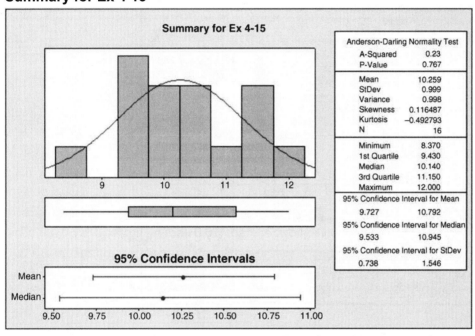

(e) Construct a 95% upper confidence interval on σ.

$\alpha = 0.05; \ \chi^2_{1-\alpha,n-1} = \chi^2_{0.95,15} = 7.2609$ (from Appendix III)

$\sigma^2 \le (n-1)S^2 / \chi^2_{1-\alpha,n-1}$

$\sigma^2 \le (16-1)0.999^2 / 7.2609$ (Equation 4.40)

$\sigma^2 \le 2.062$

$\sigma \le 1.436$

14.15. continued

(f) Does the assumption of normality seem reasonable for the output voltage?

MTB > Graph > Probability Plot > Single and ensure Distribution = Normal
Probability Plot of Ex 4-15

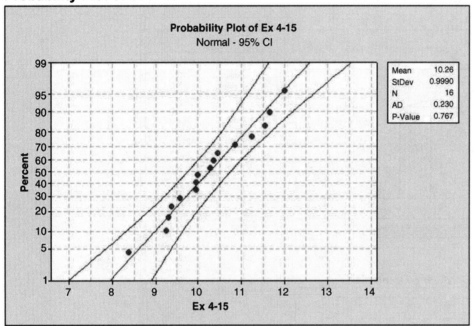

Looking at the plot, the plotted points fall approximately along a straight line, so the assumption of a normal distribution for output voltage seems appropriate.

4.17.

Two quality control technicians measured the surface finish of a metal part, obtaining the data in Table 4E.1. Assume that the measurements are normally distributed.

■ **TABLE 4E.1**
Surface Finish Data for Exercise 4.17

Technician 1	Technician 2
1.45	1.54
1.37	1.41
1.21	1.56
1.54	1.37
1.48	1.20
1.29	1.31
1.34	1.27
	1.35

(a) Test the hypothesis that the mean surface finish measurements made by the two technicians are equal. Use $\alpha = 0.05$ and assume equal variances.

Test H_0: $\mu_1 - \mu_2 = 0$ vs. H_1: $\mu_1 - \mu_2 \neq 0$. Reject H_0 if $|t_0| > t_{\alpha/2,\, n1+n2-2}$.

$$s_P = \sqrt{\frac{(n_1-1)s_1^2 + (n_2-1)s_2^2}{n_1+n_2-2}} = \sqrt{\frac{(7-1)0.115^2 + (8-1)0.125^2}{7+8-2}} = 0.1204$$
(Equations 4.51 and 4.52)

$$t_0 = \left(\bar{x}_1 - \bar{x}_2\right)\Big/\left(s_P\sqrt{1/n_1 + 1/n_2}\right) = \left(1.383 - 1.376\right)\Big/\left(0.1204\sqrt{1/7 + 1/8}\right) = 0.11$$

$t_{\alpha/2,\, n1+n2-2} = t_{0.025,\, 13} = 2.160$ (from Appendix IV)

Do not reject H_0, and conclude that there is not sufficient evidence of a difference between measurements obtained by the two technicians.

MTB > Stat > Basic Statistics > 2-Sample t
Two-Sample T-Test and CI: Ex4-17T1, Ex4-17T2

```
Two-sample T for Ex4-17T1 vs Ex4-17T2

          N   Mean   StDev   SE Mean
Ex4-17T1  7   1.383  0.115    0.043
Ex4-17T2  8   1.376  0.125    0.044

Difference = mu (Ex4-17T1) - mu (Ex4-17T2)
Estimate for difference:   0.0066
95% CI for difference:    (-0.1280, 0.1412)
T-Test of difference = 0 (vs not =): T-Value = 0.11   P-Value = 0.917   DF = 13
Both use Pooled StDev = 0.1204
```

4.17. continued

(b) What are the practical implications of the test in part (a)? Discuss what practical conclusions you would draw if the null hypothesis were rejected.

The practical implication of this test is that it does not matter which technician measures parts; the readings will be the same. If the null hypothesis had been rejected, we would have been concerned that the technicians obtained different measurements, and an investigation should be undertaken to understand why.

(c) Assuming that the variances are equal, construct a 95% confidence interval on the mean difference in surface-finish measurements.

$n_1 = 7$, $\overline{x}_1 = 1.383$, $s_1 = 0.115$; $n_2 = 8$, $\overline{x}_2 = 1.376$, $s_2 = 0.125$, $s_P = 0.120$

$\alpha = 0.05$, $t_{\alpha/2,\, n1+n2-2} = t_{0.025,\, 13} = 2.160$

$$(\overline{x}_1 - \overline{x}_2) - t_{\alpha/2, n_1 + n_2 - 2} s_p \sqrt{1/n_1 + 1/n_2} \le (\mu_1 - \mu_2) \le (\overline{x}_1 - \overline{x}_2) + t_{\alpha/2, n_1 + n_2 - 2} s_p \sqrt{1/n_1 + 1/n_2}$$

$$(1.383 - 1.376) - 2.1604(0.120)\sqrt{1/7 + 1/8} \le (\mu_1 - \mu_2) \le (1.383 - 1.376) + 2.1604(0.120)\sqrt{1/7 + 1/8}$$

$$-0.127 \le (\mu_1 - \mu_2) \le 0.141$$

(Equation 4.56)

The confidence interval for the difference contains zero. We can conclude that there is not sufficient evidence of a difference in measurements obtained by the two technicians.

(d) Test the hypothesis that the variances of the measurements made by the two technicians are equal. Use What are the practical implications if the null hypothesis is rejected?

Test $H_0 : \sigma_1^2 = \sigma_2^2$ versus $H_1 : \sigma_1^2 \ne \sigma_2^2$.

Reject H_0 if $F_0 > F_{\alpha/2, n_1 - 1, n_2 - 1}$ or $F_0 < F_{1-\alpha/2, n_1 - 1, n_2 - 1}$.

$F_0 = S_1^2 / S_2^2 = 0.115^2 / 0.125^2 = 0.8464$

$F_{\alpha/2, n_1 - 1, n_2 - 1} = F_{0.05/2, 7-1, 8-1} = F_{0.025, 6, 7} = 5.119$

$F_{1-\alpha/2, n_1 - 1, n_2 - 1} = F_{1-0.05/2, 7-1, 8-1} = F_{0.975, 6, 7} = 0.176$

4.17.(d) continued

```
MTB > Stat > Basic Statistics > 2 Variances
Test and CI for Two Variances: Ex4-17T1, Ex4-17T2

Method
Null hypothesis        Sigma(Ex4-17T1) / Sigma(Ex4-17T2) = 1
Alternative hypothesis Sigma(Ex4-17T1) / Sigma(Ex4-17T2) not = 1
Significance level     Alpha = 0.05

Statistics
Variable    N   StDev   Variance
Ex4-17T1    7   0.115    0.013
Ex4-17T2    8   0.125    0.016

Ratio of standard deviations = 0.920
Ratio of variances = 0.846
95% Confidence Intervals
                                CI for
Distribution    CI for StDev    Variance
of Data            Ratio          Ratio
Normal          (0.406, 2.195)  (0.165, 4.816)
Continuous      (0.389, 3.092)  (0.151, 9.562)

Tests
                                    Test
Method                      DF1  DF2  Statistic  P-Value
F Test (normal)              6    7     0.85      0.854
Levene's Test (any continuous) 1  13     0.01      0.920
```

Do not reject H_0, and conclude that there is not sufficient evidence of a difference in variability between measurements obtained by the two technicians. If the null hypothesis is rejected, we would have been concerned about the difference in measurement variability between the technicians, and an investigation should be undertaken to understand why.

4.17. continued

(e) Construct a 95% confidence interval estimate of the ratio of the variances of technician measurement error.

$$F_{1-\alpha/2,n_2-1,n_1-1} = F_{0.975,7,6} = 0.1954; \quad F_{\alpha/2,n_2-1,n_1-1} = F_{0.025,7,6} = 5.6955$$

$$\frac{S_1^2}{S_2^2}F_{1-\alpha/2,n_2-1,n_1-1} \leq \frac{\sigma_1^2}{\sigma_2^2} \leq \frac{S_1^2}{S_2^2}F_{\alpha/2,n_2-1,n_1-1}$$

$$\frac{0.115^2}{0.125^2}(0.1954) \leq \frac{\sigma_1^2}{\sigma_2^2} \leq \frac{0.115^2}{0.125^2}(5.6955)$$

$$0.165 \leq \frac{\sigma_1^2}{\sigma_2^2} \leq 4.821$$

(f) Construct a 95% confidence interval on the variance of measurement error for technician 2.

$$n_2 = 8; \quad \bar{x}_2 = 1.376; \quad S_2 = 0.125$$

$$\alpha = 0.05; \quad \chi^2_{\alpha/2,n_2-1} = \chi^2_{0.025,7} = 16.0128; \quad \chi^2_{1-\alpha/2,n_2-1} = \chi^2_{0.975,7} = 1.6899$$

$$\frac{(n-1)S^2}{\chi^2_{\alpha/2,n-1}} \leq \sigma^2 \leq \frac{(n-1)S^2}{\chi^2_{1-\alpha/2,n-1}}$$

$$\frac{(8-1)0.125^2}{16.0128} \leq \sigma^2 \leq \frac{(8-1)0.125^2}{1.6899}$$

$$0.007 \leq \sigma^2 \leq 0.065$$

4.17. continued

(g) Does the normality assumption seem reasonable for the data?

MTB > Graph > Probability Plot > Multiple and ensure Distribution = Normal
Probability Plot of Ex4-17T1, Ex4-17T2

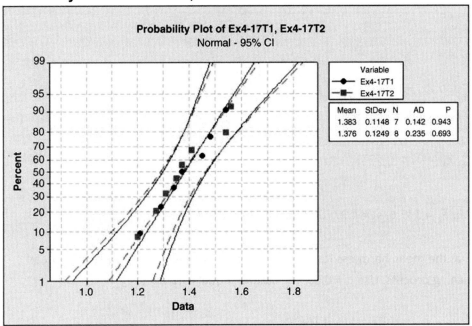

The normality assumption seems reasonable for both of these readings.

4.19.

Two different hardening processes—(1) saltwater quenching and (2) oil quenching—are used on samples of a particular type of metal alloy. The results are shown in Table 4E.2. Assume that hardness is normally distributed.

■ **TABLE 4E.2**
Hardness Data for Exercise 4.19

Saltwater Quench	Oil Quench
145	152
150	150
153	147
148	155
141	140
152	146
146	158
154	152
139	151
148	143

Saltwater quench: $n_1 = 10$, $\bar{x}_1 = 147.6$, $s_1 = 4.97$; Oil quench: $n_2 = 10$, $\bar{x}_2 = 149.4$, $s_2 = 5.46$

(a) Test the hypothesis that the mean hardness for the saltwater quenching process equals the mean hardness for the oil quenching process. Use $\alpha = 0.05$ and assume equal variances.

Test $H_0: \mu_1 - \mu_2 = 0$ vs. $H_1: \mu_1 - \mu_2 \neq 0$. Reject H_0 if $|t_0| > t_{\alpha/2,\, n1+n2-2}$.

$$s_p = \sqrt{\frac{(n_1-1)s_1^2 + (n_2-1)s_2^2}{n_1+n_2-2}} = \sqrt{\frac{(10-1)4.97^2 + (10-1)5.46^2}{10+10-2}} = 5.2217$$

(Eqns 4.51 and 4.52)

$$t_0 = \left(\bar{x}_1 - \bar{x}_2\right)\Big/\left(s_p\sqrt{1/n_1 + 1/n_2}\right) = \left(147.60 - 149.40\right)\Big/\left(5.2217\sqrt{1/10 + 1/10}\right) = -0.77$$

$t_{\alpha/2,\, n1+n2-2} = t_{0.025,\, 18} = 2.101$ (from Appendix IV)

Do not reject H_0, and conclude that there is not sufficient evidence of a difference between measurements produced by the two hardening processes.

4.19.(a) continued

MTB > Stat > Basic Statistics > 2-Sample t
Two-Sample T-Test and CI: Ex4-19SQ, Ex4-19OQ

```
Two-sample T for Ex4-19SQ vs Ex4-19OQ

           N    Mean   StDev   SE Mean
Ex4-19SQ  10   147.60   4.97     1.6
Ex4-19OQ  10   149.40   5.46     1.7

Difference = mu (Ex4-19SQ) - mu (Ex4-19OQ)
Estimate for difference:  -1.80
95% CI for difference:  (-6.71, 3.11)
T-Test of difference = 0 (vs not =): T-Value = -0.77  P-Value = 0.451  DF = 18
Both use Pooled StDev = 5.2217
```

(b) Assuming that the variances σ_1^2 and σ_2^2 are equal, construct a 95% confidence interval on the difference in mean hardness.

$\alpha = 0.05$, $t_{\alpha/2,\, n1+n2-2} = t_{0.025,\, 18} = 2.1009$

$$(\bar{x}_1 - \bar{x}_2) - t_{\alpha/2, n_1 + n_2 - 2} S_p \sqrt{1/n_1 + 1/n_2} \le (\mu_1 - \mu_2) \le (\bar{x}_1 - \bar{x}_2) + t_{\alpha/2, n_1 + n_2 - 2} S_p \sqrt{1/n_1 + 1/n_2}$$

$$(147.6 - 149.4) - 2.1009(5.22)\sqrt{1/10 + 1/10} \le (\mu_1 - \mu_2) \le (147.6 - 149.4) + 2.1009(5.22)\sqrt{1/10 + 1/10}$$

$$-6.7 \le (\mu_1 - \mu_2) \le 3.1$$

(Eqn 4.56)

The confidence interval for the difference contains zero. We conclude that there is not sufficient evidence of a difference between measurements produced by the two hardening processes.

4.19. continued

(c) Construct a 95% confidence interval on the ratio σ_1^2 / σ_2^2. Does the assumption made earlier of equal variances seem reasonable?

$$F_{1-\alpha/2,n_2-1,n_1-1} = F_{0.975,9,9} = 0.2484; \quad F_{\alpha/2,n_2-1,n_1-1} = F_{0.025,9,9} = 4.0260$$

$$\frac{S_1^2}{S_2^2}F_{1-\alpha/2,n_2-1,n_1-1} \le \frac{\sigma_1^2}{\sigma_2^2} \le \frac{S_1^2}{S_2^2}F_{\alpha/2,n_2-1,n_1-1}$$

$$\frac{4.97^2}{5.46^2}(0.2484) \le \frac{\sigma_1^2}{\sigma_2^2} \le \frac{4.97^2}{5.46^2}(4.0260)$$

$$0.21 \le \frac{\sigma_1^2}{\sigma_2^2} \le 3.34$$

```
MTB > Stat > Basic Statistics > 2 Variances
Test and CI for Two Variances: Ex4-19SQ, Ex4-19OQ

Method

Null hypothesis         Sigma(Ex4-19SQ) / Sigma(Ex4-19OQ) = 1
Alternative hypothesis  Sigma(Ex4-19SQ) / Sigma(Ex4-19OQ) not = 1
Significance level      Alpha = 0.05

Statistics
Variable    N   StDev   Variance
Ex4-19SQ   10   4.971    24.711
Ex4-19OQ   10   5.461    29.822

Ratio of standard deviations = 0.910
Ratio of variances = 0.829

95% Confidence Intervals

                                CI for
Distribution    CI for StDev    Variance
of Data            Ratio          Ratio
Normal          (0.454, 1.826)  (0.206, 3.336)
Continuous      (0.390, 2.157)  (0.152, 4.652)

Tests
                                        Test
Method                      DF1  DF2  Statistic  P-Value
F Test (normal)               9    9     0.83      0.784
Levene's Test (any continuous)  1   18     0.08      0.783
```

Since the confidence interval includes the ratio of 1, the assumption of equal variances seems reasonable.

4.19. continued

(d) Does the assumption of normality seem appropriate for these data?

MTB > Graph > Probability Plot > Multiple and ensure Distribution = Normal
Probability Plot of Ex4-19SQ, Ex4-19OQ

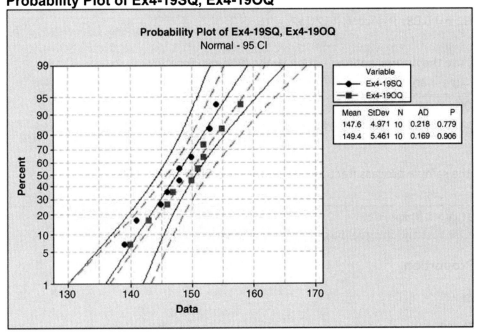

The normal distribution assumptions for both the saltwater and oil quench methods seem reasonable. However, the slopes on the normal probability plots do not appear to be the same, so the assumption of equal variances does not seem reasonable.

4.21.

A random sample of 500 connecting rod pins contains 65 nonconforming units. Estimate the process fraction nonconforming.

(a) Test the hypothesis that the true fraction defective in this process is 0.08. Use $\alpha = 0.05$.

Test $H_0: p = 0.08$ versus $H_1: p \neq 0.08$. Reject H_0 if $|Z_0| > Z_{\alpha/2}$.

$np_0 = 500(0.08) = 40$

Since $(x = 65) > (np_0 = 40)$, use the normal approximation to the binomial for $x > np_0$.

$$Z_0 = \frac{(x - 0.5) - np_0}{\sqrt{np_0(1 - p_0)}} = \frac{(65 - 0.5) - 40}{\sqrt{40(1 - 0.08)}} = 4.0387$$

$Z_{\alpha/2} = Z_{0.05/2} = Z_{0.025} = 1.96$

Reject H_0, and conclude the sample process fraction nonconforming differs from 0.08.

MTB > Stat > Basic Statistics > 1 Proportion
Under Option, select to Use test and interval based on normal distribution

Test and CI for One Proportion

```
Test of p = 0.08 vs p not = 0.08

Sample    X    N   Sample p          95% CI          Z-Value   P-Value
1        65  500   0.130000  (0.100522, 0.159478)     4.12     0.000

Using the normal approximation.
```

Note that MINITAB does not use the continuity correction factor (+/- 0.5 in Eqn 4.43) in computing the test statistic.

4.21. continued

(b) Find the P-value for this test.

Find the corresponding cumulative standard normal, $\Phi(Z_0)$ for $Z_0 = 4.0387$, and calculate:

P-value $= 2[1 - \Phi|Z_0|] = 2[1 - \Phi|4.0387|] = 2[1 - 0.99997] = 0.00006$

(c) Construct a 95% upper confidence interval on the true process fraction nonconforming.

$\alpha = 0.05$, $Z_\alpha = Z_{0.05} = 1.645$

$p \le \hat{p} + Z_\alpha \sqrt{\hat{p}(1 - \hat{p})/n}$

$p \le 0.13 + 1.645\sqrt{0.13(1 - 0.13)/500}$

$p \le 0.155$

The 95% upper confidence interval for the process fraction nonconforming is 0.155.

4.23.

A new purification unit is installed in a chemical process. Before its installation, a random sample yielded the following data about the percentage of impurity: $\bar{x}_1^2 = 9.85, s_1^2 = 6.79$ and $n_1 = 10$. After installation, a random sample resulted in $\bar{x}_2^2 = 8.08, s_2^2 = 6.18$ and $n_2 = 8$.

(a) Can you conclude that the two variances are equal? Use $\alpha = 0.05$.

Test $H_0 : \sigma_1^2 = \sigma_2^2$ versus $H_1 : \sigma_1^2 \ne \sigma_2^2$, at $\alpha = 0.05$

Reject H_0 if $F_0 > F_{\alpha/2, n_1-1, n_2-2}$ or $F_0 < F_{1-\alpha/2, n_1-1, n_2-1}$

$F_{\alpha/2, n_1-1, n_2-2} = F_{0.025,9,7} = 4.8232;\ \ F_{1-\alpha/2, n_1-1, n_2-1} = F_{0.975,9,7} = 0.2383$

$F_0 = S_1^2/S_2^2 = 6.79/6.18 = 1.0987$

$F_0 = 1.0987 < 4.8232$ and > 0.2383, so do not reject H_0

$\sqrt{s_1^2} = \sqrt{6.79} = 2.606;\ \ \sqrt{s_2^2} = \sqrt{6.18} = 2.486$

4.23.(a) continued

```
MTB > Stat > Basic Statistics > 2 Variances
Test and CI for Two Variances

Method

Null hypothesis        Sigma(1) / Sigma(2) = 1
Alternative hypothesis Sigma(1) / Sigma(2) not = 1
Significance level     Alpha = 0.05

Statistics
Sample   N  StDev  Variance
1       10  2.606    6.790
2        8  2.486    6.180

Ratio of standard deviations = 1.048
Ratio of variances = 1.099

95% Confidence Intervals
                                 CI for
Distribution   CI for StDev      Variance
of Data           Ratio           Ratio
Normal         (0.477, 2.147)  (0.228, 4.611)

Tests
                               Test
Method           DF1  DF2  Statistic  P-Value
F Test (normal)    9    7       1.10    0.922
```

The impurity variances before and after installation are the same.

4.23. continued

(b) Can you conclude that the new purification device has reduced the mean percentage of impurity? Use $\alpha = 0.05$.

Test H_0: $\mu_1 = \mu_2$ versus H_1: $\mu_1 > \mu_2$, $\alpha = 0.05$.

Reject H_0 if $t_0 > t_{\alpha, n1+n2-2}$.

$t_{\alpha, n1+n2-2} = t_{0.05, 10+8-2} = 1.746$

$$S_p = \sqrt{\frac{(n_1-1)S_1^2 + (n_2-1)S_2^2}{n_1 + n_2 - 2}} = \sqrt{\frac{(10-1)6.79 + (8-1)6.18}{10+8-2}} = 2.554$$

$$t_0 = \frac{\bar{x}_1 - \bar{x}_2}{S_p\sqrt{1/n_1 + 1/n_2}} = \frac{9.85 - 8.08}{2.554\sqrt{1/10 + 1/8}} = 1.461$$

The mean impurity after installation of the new purification unit is not less than before.

```
MTB > Stat > Basic Statistics > 2-Sample t
Two-Sample T-Test and CI

Sample   N   Mean   StDev   SE Mean
1       10   9.85   2.61     0.82
2        8   8.08   2.49     0.88

Difference = mu (1) - mu (2)
Estimate for difference:   1.77
95% lower bound for difference:   -0.35
T-Test of difference = 0 (vs >): T-Value = 1.46   P-Value = 0.082   DF = 16
Both use Pooled StDev = 2.5542
```

4.25.

The diameter of a metal rod is measured by 12 inspectors, each using both a micrometer caliper and a vernier caliper. The results are shown in Table 4E.3. Is there a difference between the mean measurements produced by the two types of caliper? Use $\alpha = 0.01$.

■ TABLE 4E.3
Measurements Made by the Inspectors for
Exercise 4.25

Inspector	Micrometer Caliper	Vernier Caliper
1	0.150	0.151
2	0.151	0.150
3	0.151	0.151
4	0.152	0.150
5	0.151	0.151
6	0.150	0.151
7	0.151	0.153
8	0.153	0.155
9	0.152	0.154
10	0.151	0.151
11	0.151	0.150
12	0.151	0.152

Test $H_0: \mu_d = 0$ versus $H_1: \mu_d \neq 0$. Reject H_0 if $|t_0| > t_{\alpha/2, n1 + n2 - 2}$.

$$\bar{d} = \frac{1}{n}\sum_{j=1}^{n}\left(x_{\text{Micrometer},j} - x_{\text{Vernier},j}\right) = \frac{1}{12}\left[(0.150 - 0.151) + \cdots + (0.151 - 0.152)\right] = -0.000417$$

$$S_d^2 = \frac{\sum_{j=1}^{n}d_j^2 - \left(\sum_{j=1}^{n}d_j\right)^2 \Big/ n}{(n-1)} = 0.001311^2$$

$$t_0 = \bar{d} \Big/ \left(S_d / \sqrt{n}\right) = -0.000417 \Big/ \left(0.001311 / \sqrt{12}\right) = -1.10$$

$$t_{\alpha/2, n1 + n2 - 2} = t_{0.005, 22} = 2.8188$$

4.25. continued

$(|t_0| = 1.10) < 2.8188$, so do not reject H_0. There is no strong evidence to indicate that the two calipers differ in their mean measurements.

```
MTB > Stat > Basic Statistics > Paired t
Paired T-Test and CI: Ex4-25MC, Ex4-25VC

Paired T for Ex4-25MC - Ex4-25VC

               N      Mean     StDev    SE Mean
Ex4-25MC      12   0.151167  0.000835  0.000241
Ex4-25VC      12   0.151583  0.001621  0.000468
Difference    12  -0.000417  0.001311  0.000379

95% CI for mean difference: (-0.001250, 0.000417)
T-Test of mean difference = 0 (vs not = 0): T-Value = -1.10  P-Value = 0.295
```

4.27.

An experiment was conducted to investigate the filling capability of packaging equipment at a winery in Newberg, Oregon. Twenty bottles of Pinot Gris were randomly selected and the fill volume (in ml) measured. Assume that fill volume has a normal distribution. The data are as follows: 753, 751, 752, 753, 753, 753, 752, 753, 754, 754, 752, 751, 752, 750, 753, 755, 753, 756, 751, and 750.

(a) Do the data support the claim that the standard deviation of fill volume is less than 1 ml? Use $\alpha = 0.05$.

$n = 20$, $\bar{x} = 752.6$ ml, $s = 1.5$, $\alpha = 0.05$

Test H_0: $\sigma^2 = 1$ versus H_1: $\sigma^2 < 1$. Reject H_0 if $\chi^2_0 < \chi^2_{1-\alpha, n-1}$.

$$\chi^2_0 = \left[(n-1)S^2\right]/\sigma^2_0 = \left[(20-1)1.5^2\right]/1 = 42.75$$

$\chi^2_{1-\alpha, n-1} = \chi^2_{0.95, 19} = 10.1170$ (from Appendix III)

$(\chi^2_0 = 42.75) > 10.1170$, so do not reject H_0. The standard deviation of the fill volume is not less than 1 ml.

4.27.(a) continued

MTB > Stat > Basic Statistics > 1 Variance
Test and CI for One Variance: Ex4-27

Method

Null hypothesis   Sigma = 1
Alternative hypothesis   Sigma < 1

The chi-square method is only for the normal distribution.
...

Statistics

Variable	N	StDev	Variance
Ex4-27	20	1.54	2.37

95% One-Sided Confidence Intervals

Variable	Method	Upper Bound for StDev	Upper Bound for Variance
Ex4-27	Chi-Square	2.11	4.44

...

Tests

Variable	Method	Test Statistic	DF	P-Value
Ex4-27	Chi-Square	44.95	19	0.999

...

4.27. continued

(b) Find a 95% two-sided confidence interval on the standard deviation of fill volume.

$\chi^2_{\alpha/2, \, n-1} = \chi^2_{0.025,19} = 32.85$; $\chi^2_{1-\alpha/2, \, n-1} = \chi^2_{0.975,19} = 8.91$ (from Appendix III)

$(n-1)S^2 / \chi^2_{\alpha/2,n-1} \leq \sigma^2 \leq (n-1)S^2 / \chi^2_{1-\alpha/2,n-1}$

$(20-1)1.5^2 / 32.85 \leq \sigma^2 \leq (20-1)1.5^2 / 8.91$

$$1.30 \leq \sigma^2 \leq 4.80$$

$$1.14 \leq \sigma \leq 2.19$$

```
MTB > Stat > Basic Statistics > 1 Variance
Test and CI for One Variance: Ex4-27

Method

Null hypothesis        Sigma = 1
Alternative hypothesis  Sigma not = 1

The chi-square method is only for the normal distribution.
...

Statistics
Variable   N   StDev  Variance
Ex4-27    20   1.54    2.37

95% Confidence Intervals
                       CI for         CI for
Variable  Method        StDev        Variance
Ex4-27    Chi-Square  (1.17, 2.25)  (1.37, 5.05)
...

Tests
                        Test
Variable  Method     Statistic  DF  P-Value
Ex4-27    Chi-Square    44.95    19   0.001
...
```

4.27.(b) continued

Also, in MINITAB:

MTB > Stat > Basic Statistics > Graphical Summary
Summary for Ex4-27

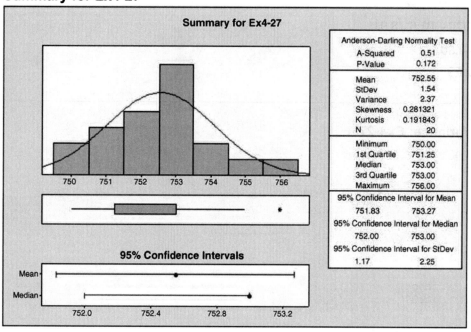

The confidence interval does not include unity; therefore, we cannot conclude that the standard deviation of fill volume is less than 1 ml.

4.27. continued

(c) Does it seem reasonable to assume that fill volume has a normal distribution?

MTB > Graph > Probability Plot > Single and ensure Distribution = Normal

Probability Plot of Ex4-27

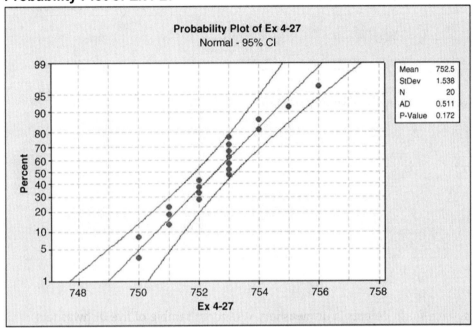

The plotted points do not fall approximately along a straight line, so the assumption that battery life is normally distributed is not appropriate.

4.29.

Consider the hypotheses

$H_0 : \mu = \mu_0$

$H_1 : \mu \neq \mu_0$

where σ^2 is known. Derive a general expression for determining the sample size for detecting a true mean of $\mu_1 \neq \mu_0$ with probability $1 - \beta$ if the type I error is α.

Let $\mu_1 = \mu_0 + \delta$.

From equation. 4.46, $\beta = \Phi\left(Z_{\alpha/2} - \delta\sqrt{n}/\sigma\right) - \Phi\left(-Z_{\alpha/2} - \delta\sqrt{n}/\sigma\right)$

If $\delta > 0$, then $\Phi\left(-Z_{\alpha/2} - \delta\sqrt{n}/\sigma\right)$ is likely to be small compared with β. So,

$\beta \approx \Phi\left(Z_{\alpha/2} - \delta\sqrt{n}/\sigma\right)$

$\Phi(\beta) \approx \Phi^{-1}\left(Z_{\alpha/2} - \delta\sqrt{n}/\sigma\right)$

$-Z_\beta \approx Z_{\alpha/2} - \delta\sqrt{n}/\sigma$

$n \approx \left[(Z_{\alpha/2} + Z_\beta)\sigma/\delta\right]^2$

4.33.

An inspector counts the surface-finish defects in dishwashers. A random sample of five dishwashers contains three such defects. Is there reason to conclude that the mean occurrence rate of surface-finish defects per dishwasher exceeds 0.5? Use the results of part (a) of Exercise 4.32 and assume that $\alpha = 0.05$.

$x \sim Poi(\lambda)$, $n = 5$, $x = 3$, $\bar{x} = x/N = 3/5 = 0.6$

Test H_0: $\lambda = 0.5$ versus H_1: $\lambda > 0.5$, at $\alpha = 0.05$. Reject H_0 if $Z_0 > Z_\alpha$.

$Z_\alpha = Z_{0.05} = 1.645$

$Z_0 = (\bar{x} - \lambda_0)/\sqrt{\lambda_0/n} = (0.6 - 0.5)/\sqrt{0.5/5} = 0.3162$

$(Z_0 = 0.3162) < 1.645$, so do not reject H_0.

4.35.

An article in *Solid State Technology* (May 1987) describes an experiment to determine the effect of C_2F_6 flow rate on etch uniformity on a silicon wafer used in integrated-circuit manufacturing. Three flow rates are tested, and the resulting uniformity (in percent) is observed for six test units at each flow rate. The data are shown in Table 4E.4.

■ TABLE 4E.4
Uniformity Data for Exercise 4.35

C_2F_6 Flow (SCCM)	Observations					
	1	2	3	4	5	6
125	2.7	2.6	4.6	3.2	3.0	3.8
160	4.6	4.9	5.0	4.2	3.6	4.2
200	4.6	2.9	3.4	3.5	4.1	5.1

(a) Does C_2F_6 flow rate affect etch uniformity? Answer this question by using an analysis of variance with $\alpha = 0.05$.

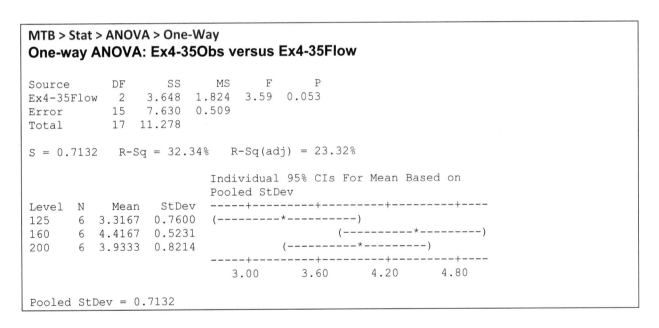

```
MTB > Stat > ANOVA > One-Way
One-way ANOVA: Ex4-35Obs versus Ex4-35Flow

Source        DF      SS     MS      F      P
Ex4-35Flow     2   3.648  1.824   3.59  0.053
Error         15   7.630  0.509
Total         17  11.278

S = 0.7132    R-Sq = 32.34%    R-Sq(adj) = 23.32%

                          Individual 95% CIs For Mean Based on
                          Pooled StDev
Level   N    Mean   StDev    -----+---------+---------+---------+----
125     6  3.3167  0.7600    (---------*----------)
160     6  4.4167  0.5231                     (----------*---------)
200     6  3.9333  0.8214              (----------*---------)
                             -----+---------+---------+---------+----
                                3.00      3.60      4.20      4.80

Pooled StDev = 0.7132
```

$(F_{0.05,2,15} = 3.6823) > (F_0 = 3.59)$, so flow rate does not affect etch uniformity at a significance level $\alpha = 0.05$. However, the *P*-value = 0.053 is just slightly greater than 0.05, so there is some evidence that gas flow rate affects the etch uniformity.

4.35. continued

(b) Construct a box plot of the etch uniformity data. Use this plot, together with the analysis of variance results, to determine which gas flow rate would be best in terms of etch uniformity (a small percentage is best).

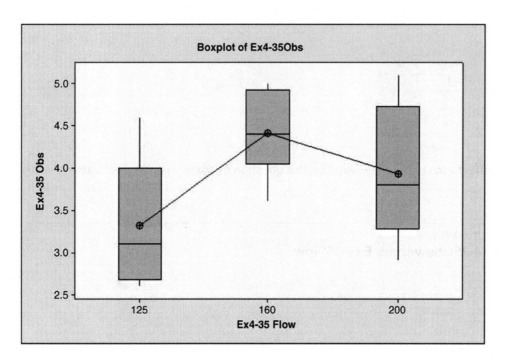

Gas flow rate of 125 SCCM gives smallest mean percentage uniformity.

4.35. continued

(c) Plot the residuals versus predicted C_2F_6 flow. Interpret this plot.

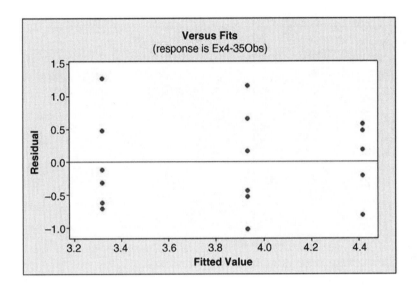

Residuals are satisfactory, with no unusual patterns and approximately equal variance over the range of fitted values.

4.35. continued

(d) Does the normality assumption seem reasonable in this problem?

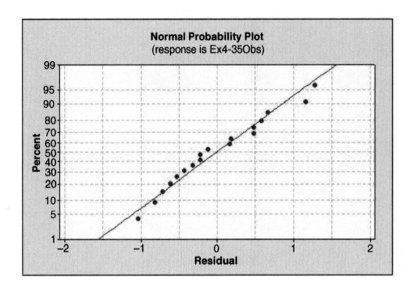

The normality assumption is reasonable.

4.37.

An article in the *ACI Materials Journal* (Vol. 84, 1987, pp. 213–216) describes several experiments investigating the rodding of concrete to remove entrapped air. A 3-in.-diameter cylinder was used, and the number of times this rod was used is the design variable. The resulting compressive strength of the concrete specimen is the response. The data are shown in Table 4E.5.

■ **TABLE 4E.5**
Compressive Strength Data for Exercise 4.37

Rodding Level	Compressive Strength		
10	1,530	1,530	1,440
15	1,610	1,650	1,500
20	1,560	1,730	1,530
25	1,500	1,490	1,510

(a) Is there any difference in compressive strength due to the rodding level? Answer this question by using the analysis of variance with $\alpha = 0.05$.

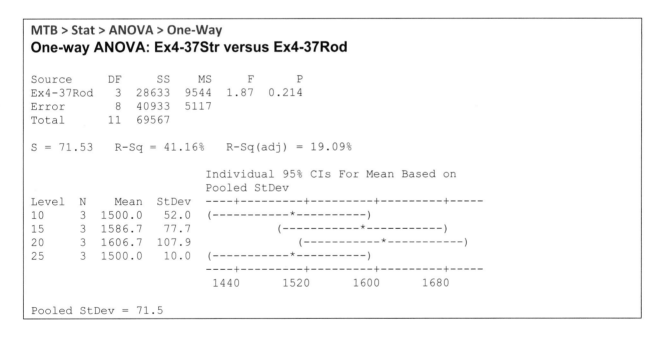

```
MTB > Stat > ANOVA > One-Way
One-way ANOVA: Ex4-37Str versus Ex4-37Rod

Source      DF     SS    MS     F      P
Ex4-37Rod    3   28633  9544  1.87  0.214
Error        8   40933  5117
Total       11   69567

S = 71.53    R-Sq = 41.16%    R-Sq(adj) = 19.09%

                            Individual 95% CIs For Mean Based on
                            Pooled StDev
Level  N    Mean   StDev    ----+---------+---------+---------+-----
10     3  1500.0    52.0    (-----------*----------)
15     3  1586.7    77.7             (-----------*-----------)
20     3  1606.7   107.9               (-----------*-----------)
25     3  1500.0    10.0    (-----------*----------)
                            ----+---------+---------+---------+-----
                            1440      1520      1600      1680

Pooled StDev = 71.5
```

ANOVA *P*-value = 0.214, so no difference due to rodding level at $\alpha = 0.05$.

4.37. continued

(b) Construct box plots of compressive strength by rodding level. Provide a practical interpretation of these plots.

Boxplot of Ex4-37Str

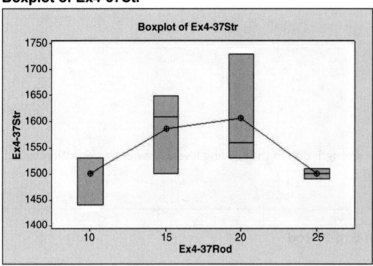

Level 25 exhibits considerably less variability than the other three levels.

4.37. continued

(c) Construct a normal probability plot of the residuals from this experiment. Does the assumption of a normal distribution for compressive strength seem reasonable?

Normplot of Residuals for Ex4-37Str

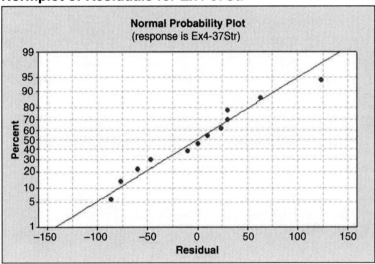

The normal distribution assumption for compressive strength is reasonable.

4.39.

An aluminum producer manufactures carbon anodes and bakes them in a ring furnace prior to use in the smelting operation. The baked density of the anode is an important quality characteristic, as it may affect anode life. One of the process engineers suspects that firing temperature in the ring furnace affects baked anode density. An experiment was run at four different temperature levels, and six anodes were baked at each temperature level. The data from the experiment are shown in Table 4E.6.

■ **TABLE 4E.6**
Baked Density Data for Exercise 4.39

Temperature (°C)	Density					
500	41.8	41.9	41.7	41.6	41.5	41.7
525	41.4	41.3	41.7	41.6	41.7	41.8
550	41.2	41.0	41.6	41.9	41.7	41.3
575	41.0	40.6	41.8	41.2	41.9	41.5

(a) Does firing temperature in the ring furnace affect mean baked anode density?

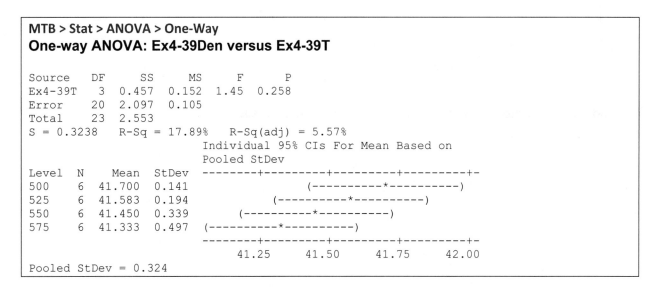

```
MTB > Stat > ANOVA > One-Way
One-way ANOVA: Ex4-39Den versus Ex4-39T

Source     DF      SS      MS      F       P
Ex4-39T     3   0.457   0.152   1.45   0.258
Error      20   2.097   0.105
Total      23   2.553
S = 0.3238    R-Sq = 17.89%    R-Sq(adj) = 5.57%
                              Individual 95% CIs For Mean Based on
                              Pooled StDev
Level   N    Mean    StDev   --------+---------+---------+---------+-
500     6   41.700   0.141                      (----------*----------)
525     6   41.583   0.194                  (----------*----------)
550     6   41.450   0.339           (----------*----------)
575     6   41.333   0.497   (----------*----------)
                              --------+---------+---------+---------+-
                                  41.25     41.50     41.75     42.00
Pooled StDev = 0.324
```

ANOVA *P*-value = 0.258. At α = 0.05, temperature level does not significantly affect mean baked anode density.

4.39. continued

(b) Find the residuals for this experiment and plot them on a normal probability scale. Comment on the plot.

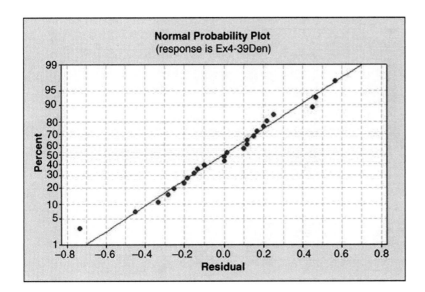

Normality assumption is reasonable.

4.39. continued

(c) What firing temperature would you recommend using?

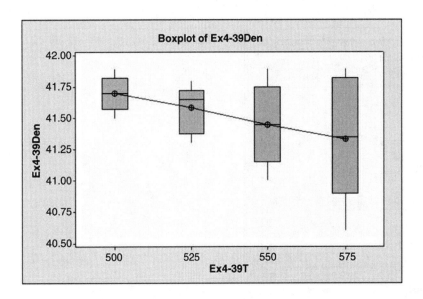

Since statistically there is no evidence to indicate that the means are different, select the temperature with the smallest variance, 500°C (see Boxplot), which probably also incurs the least cost (lowest temperature).

4.41.

An article in Environmental International (Vol. 18, No. 4, 1992) describes an experiment in which the amount of radon released in showers was investigated. Radon-enriched water was used in the experiment, and six different orifice diameters were tested in showerheads. The data from the experiment are shown in Table 4E.7.

■ TABLE 4E.7
Radon Data for the Experiment in Exercise 4.41

Orifice Diameter	Radon Released (%)			
0.37	80	83	83	85
0.51	75	75	79	79
0.71	74	73	76	77
1.02	67	72	74	74
1.40	62	62	67	69
1.99	60	61	64	66

(a) Does the size of the orifice affect the mean percentage of radon released? Use the analysis of variance and $\alpha = 0.05$.

```
MTB > Stat > ANOVA > One-Way
One-way ANOVA: Ex4-41Rad versus Ex4-41Dia

Source       DF       SS      MS      F      P
Ex4-41Dia     5  1133.38  226.68  30.85  0.000
Error        18   132.25    7.35
Total        23  1265.63

S = 2.711   R-Sq = 89.55%   R-Sq(adj) = 86.65%

                            Individual 95% CIs For Mean Based on
                            Pooled StDev
Level  N    Mean   StDev  ----+---------+---------+---------+-----
0.37   4  82.750   2.062                              (---*---)
0.51   4  77.000   2.309                    (---*---)
0.71   4  75.000   1.826                  (---*---)
1.02   4  71.750   3.304              (----*---)
1.40   4  65.000   3.559      (---*---)
1.99   4  62.750   2.754  (---*---)
                           ----+---------+---------+---------+-----
                           63.0      70.0      77.0      84.0

Pooled StDev = 2.711
```

Orifice size does affect mean % radon release, at $\alpha = 0.05$.

4.41.(a) continued

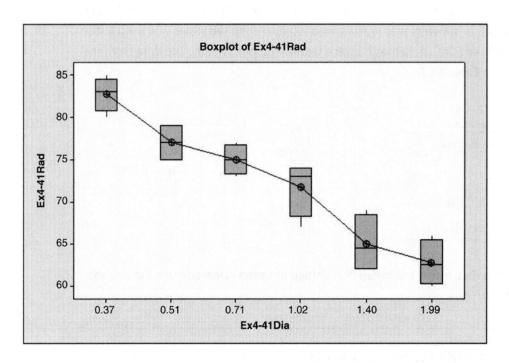

Smallest % radon released at 1.99 and 1.4 orifice diameters.

4.41. continued

(b) Analyze the residuals from this experiment.

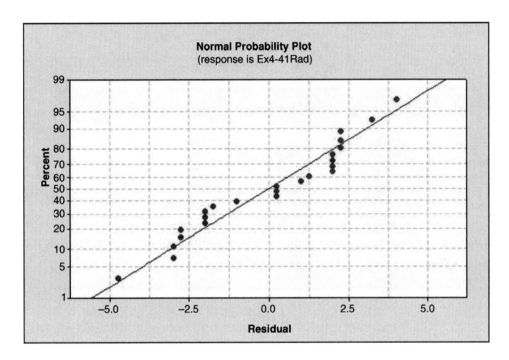

Residuals violate the assumption of normally distributed errors.

4.41.(b) continued

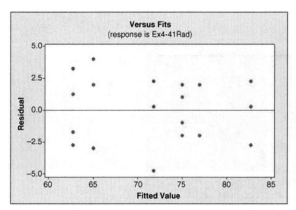

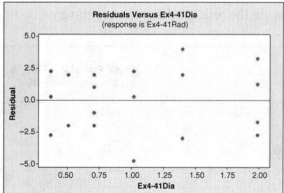

Variability in residuals does not appear to depend on the magnitude of predicted (or fitted) values.

The assumption of equal variance at each factor level appears to be violated, with larger variances at the larger diameters (1.02, 1.40, 1.99).

4.53.

Consider the Minitab output below.

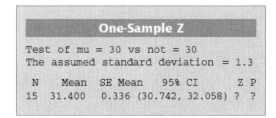

(a) Fill in the missing values. What conclusions would you draw?

$$Z = \frac{31.400 - 30}{0.336} = 4.1667$$
$$P = 2\Pr\left[Z > 4.1667\right] = 0.00003$$

The small *P*-value would lead to concluding that the mean is not 30.

4.53. continued

(b) Is this a one-sided or two-sided test?

This is a two-sided test (given in the first line of the output). This is why the *P*-value above was double
the probability instead of just simply the probability.

(c) Use the output and a normal table to find a 95% CI on the mean.

$$\bar{x} - Z_{\alpha/2} \frac{\sigma}{\sqrt{n}} \le \mu \le \bar{x} - Z_{\alpha/2} \frac{\sigma}{\sqrt{n}}$$

$$31.4 - 1.96 \frac{1.3}{\sqrt{15}} \le \mu \le 31.4 + 1.96 \frac{1.3}{\sqrt{15}} \quad \text{(Equation 4.29)}$$

$$30.7421 \le \mu \le 32.0579$$

This answer agrees with the given CI.

(d) How was the SE mean calculated?

$$\text{SE mean} = \frac{1.3}{\sqrt{15}} = 0.336 \,.$$

(e) What is the P-value if the alternative hypothesis is H_1: $\mu > 30$?

P-value=0.00003/2=0.000015

4.55.

Suppose that you are testing $H_0 : \mu_1 = \mu_2$ versus $H_1 : \mu_1 \neq \mu_2$ with a $n_1 = n_2 = 10$. Use the table of the t distribution percentage points of find lower and upper bounds on the P-value of the following observed values of the test statistic:

(a) $t_0 = 2.48$

t (9)	Prob
2.262	0.025
2.48	p-val
2.821	0.01

Since this is a two-sided test, we double the probabilities. So, $0.02 < P$-value < 0.05.

(b) $t_0 = -2.41$ Notice that we can find the probabilities associated with the absolute value of the test statistic, since the t distribution is symmetric about 0.

t (9)	Prob
2.262	0.025
2.41	p-val
2.821	0.01

Since this is a two-sided test, we double the probabilities. So, $0.02 < P$-value < 0.05.

4.55. continued

(c) $t_0 = 2.98$

$t(9)$	Prob
2.821	0.01
2.98	p-val
3.250	0.005

Since this is a two-sided test, we double the probabilities. So, $0.01 < P\text{-value} < 0.02$.

(d) $t_0 = 1.89$

$t(9)$	Prob
1.833	0.05
1.89	p-val
2.262	0.025

Since this is a two-sided test, we double the probabilities. So, $0.05 < P\text{-value} < 0.10$.

4.57.

Consider the Minitab output below.

Test and CI for One Proportion						
Test of p = 0.3 vs p not = 0.3						
Sample	X	N	Sample p	95% CI	Z-Value	P-Value
1	98	300	0.326667	(0.273596, 0.379737)	1.01	0.313

(a) Is this a one-sided or two-sided test?

This is a two-sided test (given in the first line of the output).

(b) Can the null hypothesis be rejected at the 0.05 level?

Since the given *P*-value of 0.313 is larger than 0.05, we cannot reject the null hypothesis at this level.

(c) Construct an approximate 90% CI for *p*.

$$\hat{p} - Z_{\alpha/2}\sqrt{\frac{\hat{p}(1-\hat{p})}{n}} \leq p \leq \hat{p} - Z_{\alpha/2}\sqrt{\frac{\hat{p}(1-\hat{p})}{n}}$$

$$0.3267 - 1.645\sqrt{\frac{0.3267(1-0.3267)}{300}} \leq p \leq 0.3267 + 1.645\sqrt{\frac{0.3267(1-0.3267)}{300}} \qquad \text{(Equation 4.44)}$$

$$0.2822 \leq p \leq 0.3712$$

(d) What is the P-value if the alternative hypothesis is $H_1: p > 0.3$?

The *P*-value is half the *P*-value for the two sided alternative, so $P = 0.131/2 = 0.156$.

4.61.

Consider the Minitab ANOVA output below. Fill in the blanks. You may give bounds on the *P*-value. What conclusions can you draw based on the information in this display?

```
                        One-Way ANOVA

Source     DF      SS      MS      F      P
Factor      3    54.91     ?      ?      ?
Error       ?    19.77     ?
Total      15    74.67

S = 1.283        R-Sq = 73.53%  R-Sq(adj) = 66.91%
```

MSFactor = SSFactor/3 = 18.30

DF Error = DF Total – DF Factor = 12

MSError = SSError/12 = 19.77/12 = 1.65

F = MSFactor/MSError = 18.30/1.65 = 11.09

P-value < 0.01

Since the P-value is smaller than 0.01, there is strong evidence against the null hypothesis. We conclude that the means are not the same for all factor levels, and that changing the level of the factor will significantly affect the response.

Chapter 5

Methods and Philosophy of Statistical Process Control

LEARNING OBJECTIVES

After completing this chapter you should be able to:

1. Understand chance and assignable causes of variability in a process
2. Explain the statistical basis of the Shewhart control chart, including choice of sample size, control limits, and sampling interval
3. Explain the rational subgroup concept
4. Understand the basic tools of SPC; the histogram or stem-and-leaf plot, the check sheet, the Pareto chart, the cause-and-effect diagram, the defect concentration diagram, the scatter diagram, and the control chart
5. Explain phase I and phase II use of control charts
6. Explain how average run length is used as a performance measure for a control chart
7. Explain how sensitizing rules and pattern recognition are used in conjunction with control charts

IMPORTANT TERMS AND CONCEPTS

Action Limits

Assignable causes of variation

Average run length (ARL)

Average time to signal

Cause-and-effect diagram

Chance causes of variation

Check sheet

Control chart

Control limits

Out-of-control process

Out-of-control-action plan (OCAP)

Pareto chart

Patterns on control charts

Phase I and phase II applications

Rational subgroups

Sample size for control charts

Sampling frequency for control charts

Scatter diagram

Defect concentration diagram

Designed experiments

Factorial experiment

Flow charts and operations process charts, and value stream mapping

In-control process

Magnificent seven

Sensitizing rules for control charts

Shewhart control charts

Statistical control of a process

Statistical process control (SPC)

Three-sigma control limits

Warning limits

EXERCISES

5.17.

Consider the control chart shown here. Does the pattern appear random?

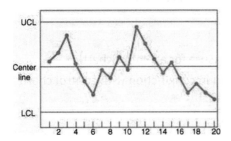

Evidence of runs, trends or cycles? NO. There are no runs of 5 points or cycles. So, we can say that the plot point pattern appears to be random.

5.19.

Consider the control chart shown here. Does the pattern appear random?

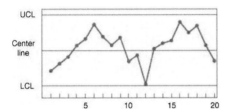

Evidence of runs, trends or cycles? YES, there is a "low - high - low - high - low" wave (all samples), which might be a cycle. So, we can say that the pattern does not appear random.

5.21.

Apply the Western Electric rules to the control chart in Exercise 5.17. Are any of the criteria for declaring the process out of control satisfied?

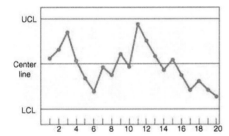

Check:

- Any point outside the 3-sigma control limits? NO.
- 2 of 3 beyond 2 sigma of centerline? NO.
- 4 of 5 at 1 sigma or beyond of centerline? YES. Points #17, 18, 19, and 20 are outside the lower 1-sigma area.
- 8 consecutive points on one side of centerline? NO.

A one out-of-control criterion is satisfied.

5.23.

Apply the Western Electric rules to the control chart presented in Exercise 5.19. Would these rules result in any out-of-control signals?

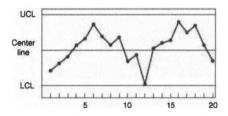

Check:

- Any point outside the 3-sigma control limits? NO. (Point #12 is within the lower 3-sigma control limit.)
- 2 of 3 beyond 2 sigma of centerline? YES, points #16, 17, and 18.
- 4 of 5 at 1 sigma or beyond of centerline? YES, points #5, 6, 7, 8, and 9.
- 8 consecutive points on one side of centerline? NO.

Two out-of-control criteria are satisfied.

Chapter 6

Control Charts for Variables

LEARNING OBJECTIVES

After completing this chapter you should be able to:

1. Understand the statistical basis of Shewhart control charts for variables
2. Know how to design variables control charts
3. Know how to set up and use $\bar{x}$ and R control charts
4. Know how to estimate process capability from the control chart information
5. Know how to interpret patterns on $\bar{x}$ and R control charts
6. Know how to set up and use $\bar{x}$ and s or s^2 control charts
7. Know how to set up and use control charts for individual measurements
8. Understand the importance of the normality assumption for individuals control charts and know how to check this information
9. Understand the rational subgroup concept for variables control charts
10. Determine the average run length for variables control charts

IMPORTANT TERMS AND CONCEPTS

Average run length

Control chart for individuals units

Control limits

Interpretation of control charts

Moving-range control chart

Natural tolerance limits for a process

Normality and control charts

Operating characteristic (OC) curve for the $\bar{x}$ control chart

Patterns on control charts

Phase I control chart usage

Phase II control chart usage

Probability limits for control charts

Process capability

Process capability ratio (PCR) C_p

R control chart

Rational subgroups

s control chart

s^2 control chart

Shewhart control charts

Specification limits

Three-sigma control limits

Tier chart or tolerance diagram

Trial control limits

Variable sample size on control charts

Variables control charts

$\bar{x}$ control chart

EXERCISES

Minitab® Notes:

1. The Minitab convention for determining whether a point is out of control is: (1) if a plot point is within the control limits, it is in control, or (2) if a plot point is on or beyond the limits, it is out of control.
2. Minitab uses pooled standard deviation to estimate standard deviation for control chart limits and capability estimates. This can be changed in dialog boxes or under Tools>Options>Control Charts and Quality Tools>Estimating Standard Deviation.
3. Minitab defines some sensitizing rules for control charts differently than the standard rules. In particular, a run of *n* consecutive points on one side of the center line is defined as 9 points, not 8. This can be changed in dialog boxes, or under Tools > Options > Control Charts and Quality Tools > Tests.

6.1.

A manufacturer of component for automobile transmissions wants to use control charts to monitor a process producing a shaft. The resulting data from 20 samples of 4 shaft diameters that have been measured are:

$$\sum_{i=1}^{20} \bar{x}_i = 10.275, \quad \sum_{i=1}^{20} R_i = 1.012$$

(a) Find the control limits that should be used on the $\bar{x}$ and R control charts.

For $n = 4$, $A_2 = 0.729$, $D_4 = 2.282$, $D_3 = 0$

$$\bar{\bar{x}} = \frac{\sum_{i=1}^{m} \bar{x}_i}{m} = \frac{10.275}{20} = 0.5138$$

$$\bar{R} = \frac{\sum_{i=1}^{m} R_i}{m} = \frac{1.012}{20} = 0.0506$$

$\text{UCL}_{\bar{x}} = \bar{\bar{x}} + A_2\bar{R} = 0.5138 + 0.729(0.0506) = 0.5507$ (Equations 6.2, 6.3, 6.4, 6.5)

$\text{CL}_{\bar{x}} = \bar{\bar{x}} = 0.5138$

$\text{LCL}_{\bar{x}} = \bar{\bar{x}} - A_2\bar{R} = 0.5138 - 0.729(0.0506) = 0.4769$

$\text{UCL}_R = D_4\bar{R} = 2.282(0.0506) = 0.1155$

$\text{CL}_R = \bar{R} = 0.0506$

$\text{LCL}_R = D_3\bar{R} = 0(0.0506) = 0.00$

6.1. continued

(b) Assume that the 20 preliminary samples plot in control on both charts. Estimate the process mean and standard deviation.

Process mean = $\bar{\bar{x}} = 0.5138$

For $n = 4$, $d_2 = 2.059$

Process standard deviation = $\hat{\sigma} = \bar{R}/d_2 = 0.0506/2.059 = 0.0246$ (Equation 6.6)

6.3.

Reconsider the situation described in Exercise 6.1. Suppose that several of the preliminary 20 samples plot out of control on the R chart. Does this have any impact on the reliability of the control limits on the $\bar{x}$ chart?

Yes. Out-of-control samples on the R chart signal a potential problem with process variability, resulting in an unreliable estimate of the process standard deviation, and consequently impacting the accuracy of the upper and lower control limits on the $\bar{x}$ chart.

6.7.

The data shown in Table 6E.2 are $\bar{x}$ and R values for 24 samples of size $n = 5$ taken from a process producing bearings. The measurements are made on the inside diameter of the bearing, with only the last three decimals recorded (i.e., 34.5 should be 0.50345).

■ TABLE 6E.2
Bearing Diameter Data

Sample Number	$\bar{x}$	R	Sample Number	$\bar{x}$	R
1	34.5	3	13	35.4	8
2	34.2	4	14	34.0	6
3	31.6	4	15	37.1	5
4	31.5	4	16	34.9	7
5	35.0	5	17	33.5	4
6	34.1	6	18	31.7	3
7	32.6	4	19	34.0	8
8	33.8	3	20	35.1	4
9	34.8	7	21	33.7	2
10	33.6	8	22	32.8	1
11	31.9	3	23	33.5	3
12	38.6	9	24	34.2	2

6.7. continued

(a) Set up $\bar{x}$ and R charts on this process. Does the process seem to be in statistical control? If necessary, revise the trial control limits.

For $n = 5$, $A_2 = 0.577$, $D_4 = 2.114$, $D_3 = 0$

$$\bar{\bar{x}} = \frac{\bar{x}_1 + \bar{x}_2 + \cdots + \bar{x}_m}{m} = \frac{34.5 + 34.2 + \cdots + 34.2}{24} = 34.00$$

$$\bar{R} = \frac{R_1 + R_2 + \cdots + R_m}{m} = \frac{3 + 4 + \cdots + 2}{24} = 4.71$$

$$UCL_{\bar{x}} = \bar{\bar{x}} + A_2\bar{R} = 34.00 + 0.577(4.71) = 36.72$$

$$CL_{\bar{x}} = \bar{\bar{x}} = 34.00 \qquad\qquad \text{(Equations 6.2, 6.3, 6.4, 6.5)}$$

$$LCL_{\bar{x}} = \bar{\bar{x}} - A_2\bar{R} = 34.00 - 0.577(4.71) = 31.29$$

$$UCL_R = D_4\bar{R} = 2.114(4.71) = 9.96$$

$$CL_R = \bar{R} = 4.71$$

$$LCL_R = D_3\bar{R} = 0(4.71) = 0.00$$

Note: Minitab does not have functionality to create control charts from pre-summarized data ($\bar{x}$'s and R's). The charts below were created in Excel.

6.7.(a) continued

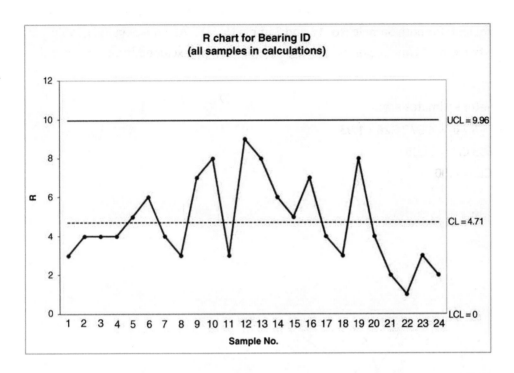

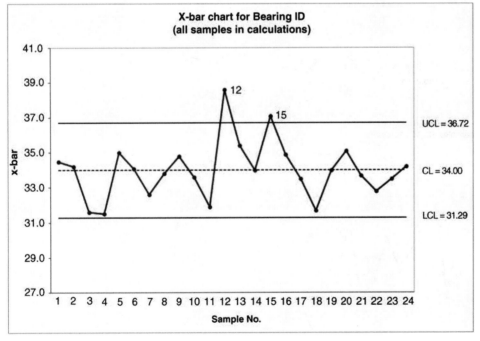

6.7.(a) continued

The R chart is in control, so the $\bar{x}$ chart can be evaluated. The process is not in statistical control; $\bar{x}$ is beyond the upper control limit for both Sample No. 12 and Sample No. 15. Assuming an assignable cause is found for these two out-of-control points, the two samples can be excluded from the control limit calculations.

The new process parameter estimates are:
$$\bar{\bar{x}} = 33.65; \quad \bar{R} = 4.5; \quad \hat{\sigma}_x = \bar{R}/d_2 = 4.5/2.326 = 1.93$$
$$UCL_{\bar{x}} = 36.25; CL_{\bar{x}} = 33.65; LCL_{\bar{x}} = 31.06$$
$$UCL_R = 9.52; CL_R = 4.5; LCL_R = 0.00$$

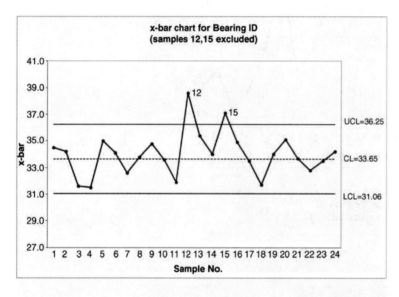

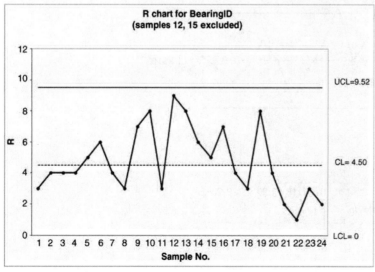

6.7. continued

(b) If specifications on this diameter are 0.5030 ± 0.0010, find the percentage of nonconforming bearings produced by this process. Assume that diameter is normally distributed.

$$\hat{p} = \Pr\{x < LSL\} + \Pr\{x > USL\} = \Pr\{x < 20\} + \Pr\{x > 40\} = \Pr\{x < 20\} + \left[1 - \Pr\{x < 40\}\right]$$

$$= \Phi\left(\frac{20 - 33.65}{1.93}\right) + \left[1 - \Phi\left(\frac{40 - 33.65}{1.93}\right)\right] \qquad \text{(p. 242)}$$

$$= \Phi(-7.07) + 1 - \Phi(3.29) = 0 + 1 - 0.99950 = 0.00050$$

6.9.

The data shown in Table 6E.4 are the deviations from nominal diameter for holes drilled din a carbon-fiber composite material used in aerospace manufacturing. The values reported are deviations from nominal in ten-thousandths of an inch.

■ TABLE 6E.4
Hole Diameter Data for Exercise 6.9

Sample Number	x_1	x_2	x_3	x_4	x_5
1	-30	+50	-20	+10	+30
2	0	+50	-60	-20	+30
3	-50	+10	+20	+30	+20
4	-10	-10	+30	-20	+50
5	+20	-40	+50	+20	+10
6	0	0	+40	-40	+20
7	0	0	+20	-20	-10
8	+70	-30	+30	-10	0
9	0	0	+20	-20	+10
10	+10	+20	+30	+10	+50
11	+40	0	+20	0	+20
12	+30	+20	+30	+10	+40
13	+30	-30	0	+10	+10
14	+30	-10	+50	-10	-30
15	+10	-10	+50	+40	0
16	0	0	+30	-10	0
17	+20	+20	+30	+30	-20
18	+10	-20	+50	+30	+10
19	+50	-10	+40	+20	0
20	+50	0	0	+30	+10

6.9. continued

(a) Set up $\bar{x}$ and R charts on the process. Is the process in statistical control?

MTB > Stat > Control Charts > Variables Charts for Subgroups > Xbar-R

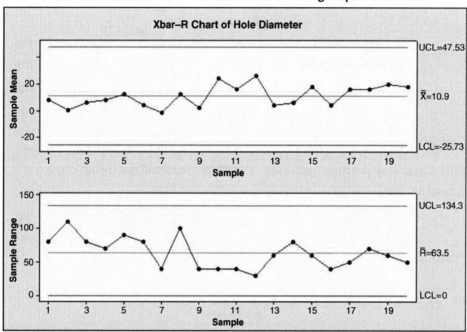

The process is in statistical control with no out-of-control signals, runs, trends, or cycles.

(b) Estimate the process standard deviation using the range method.

$$\hat{\sigma}_x = \bar{R}/d_2 = 63.5/2.326 = 27.3$$

6.9. continued

(c) If specifications are at nominal ± 100, what can you say about the capability of this process? Calculate the PCR C_p.

USL = +100, LSL = −100

$$\hat{C}_p = \frac{USL - LSL}{6\hat{\sigma}_x} = \frac{+100 - (-100)}{6(27.3)} = 1.22 \text{, so the process is capable.}$$

MTB > Stat > Quality Tools > Capability Analysis > Normal

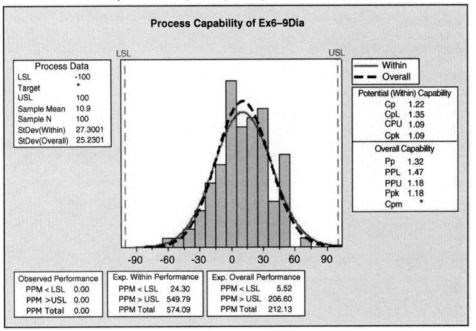

6.11.

The fill volume of soft-drink beverage bottles is an important quality characteristic. The colume is measured (approximately) by placing a gauge over the crown and comparing the height of the liquid in the neck of the bottle against a coded scale. On this scale a reading fo zero corresponds to the correct fill height. Fifteen samples of size $n = 10$ have been analyzed, and the fill heights are shown in Table 6E.6.

■ **TABLE 6E.6**
Fill Height Data for Exercise 6.11

Sample Number	x_1	x_2	x_3	x_4	x_5	x_6	x_7	x_8	x_9	x_{10}
1	2.5	0.5	2.0	−1.0	1.0	−1.0	0.5	1.5	0.5	−1.5
2	0.0	0.0	0.5	1.0	1.5	1.0	−1.0	1.0	1.5	−1.0
3	1.5	1.0	1.0	−1.0	0.0	−1.5	−1.0	−1.0	1.0	−1.0
4	0.0	0.5	−2.0	0.0	−1.0	1.5	−1.5	0.0	−2.0	−1.5
5	0.0	0.0	0.0	−0.5	0.5	1.0	−0.5	−0.5	0.0	0.0
6	1.0	−0.5	0.0	0.0	0.0	0.5	−1.0	1.0	−2.0	1.0
7	1.0	−1.0	−1.0	−1.0	0.0	1.5	0.0	1.0	0.0	0.0
8	0.0	−1.5	−0.5	1.5	0.0	0.0	0.0	−1.0	0.5	−0.5
9	−2.0	−1.5	1.5	1.5	0.0	0.0	0.5	1.0	0.0	1.0
10	−0.5	3.5	0.0	−1.0	−1.5	−1.5	−1.0	−1.0	1.0	0.5
11	0.0	1.5	0.0	0.0	2.0	−1.5	0.5	−0.5	2.0	−1.0
12	0.0	−2.0	−0.5	0.0	−0.5	2.0	1.5	0.0	0.5	−1.0
13	−1.0	−0.5	−0.5	−1.0	0.0	0.5	0.5	−1.5	−1.0	−1.0
14	0.5	1.0	−1.0	−0.5	−2.0	−1.0	−1.5	0.0	1.5	1.5
15	1.0	0.0	1.5	1.5	1.0	−1.0	0.0	1.0	−2.0	−1.5

6.11. continued

(a) Set up $\bar{x}$ and S control charts on this process. Does the process exhibit statistical control? If necessary construct revised control limits.

MTB > Stat > Control Charts > Variables Charts for Subgroups > Xbar-S
Under "Options, Estimate" select Sbar as method to estimate standard deviation.

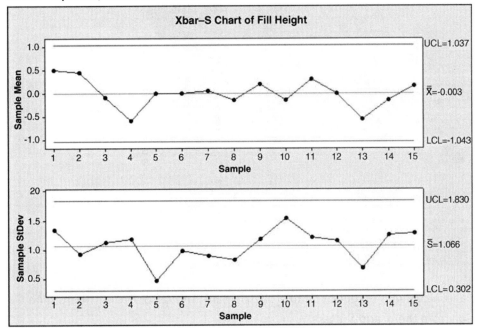

The process is in statistical control, with no out-of-control signals, runs, trends, or cycles.

6.11. continued

(b) Set up an R chart, and compare it with the s chart in part (a).

MTB > Stat > Control Charts > Variables Charts for Subgroups > R

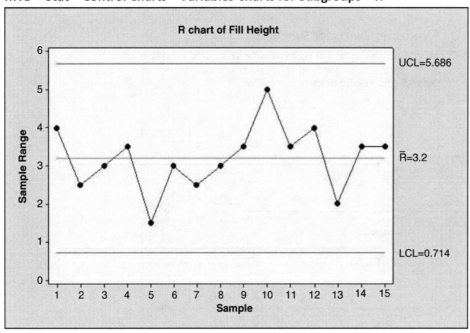

The process is in statistical control, with no out-of-control signals, runs, trends, or cycles. There is no difference in interpretation from the $\bar{x}$ and S chart.

(c) Set up an s^2 chart and compare it with the s chart in part (a).

Let $\alpha = 0.010$. $n = 15$, $\bar{s} = 1.066$.

$CL = \bar{s}^2 = 1.066^2 = 1.136$

$UCL = \bar{s}^2 / (n-1) \, \chi^2_{\alpha/2, n-1} = 1.066^2 / (15-1) \left(\chi^2_{0.010/2, 15-1} \right) = 1.066^2 / (15-1)(31.32) = 2.542$

$LCL = \bar{s}^2 / (n-1) \, \chi^2_{1-(\alpha/2), n-1} = 1.066^2 / (15-1) \left(\chi^2_{1-(0.010/2), 15-1} \right) = 1.066^2 / (15-1)(4.07) = 0.330$

6.11.(c) continued

Minitab's control chart options do not include an s^2 or variance chart. To construct an s^2 control chart, first calculate the sample standard deviations and then create a time series plot. To obtain sample standard deviations: **Stat > Basic Statistics > Store Descriptive Statistics**. "Variables" is column with sample data (Ex6.5Vol), and "By Variables" is the sample ID column (Ex6.5Sample). In "Statistics" select "Variance". Results are displayed in the session window. Copy results from the session window by holding down the keyboard "Alt" key, selecting only the variance column, and then copying & pasting to an empty worksheet column (results in Ex6.5Variance).

Graph > Time Series Plot > Simple
Control limits can be added using: Time/Scale > Reference Lines > Y positions

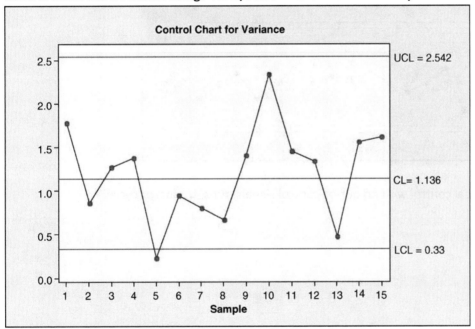

Sample 5 signals out of control below the lower control limit. Otherwise there are no runs, trends, or cycles. If the limits had been calculated using $\alpha = 0.0027$ (not tabulated in textbook), sample 5 would be within the limits, and there would be no difference in interpretation from either the $\bar{x} - s$ or the $x-R$ chart.

6.13.

Rework Exercise 6.8 using the *s* chart.

MTB > Stat > Control Charts > Variables Charts for Subgroups > Xbar-S

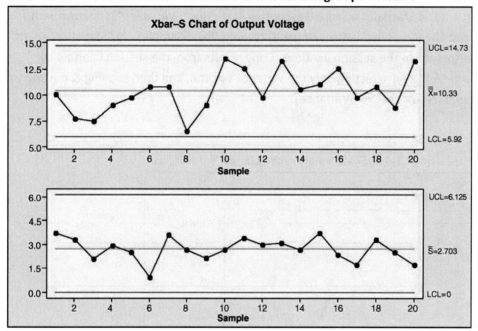

The process is in statistical control with no out-of-control signals, runs, trends, or cycles.

6.15.

Consider the piston ring data shown in Table 6.3. Assume that the specifications on this component are 74.000 ± 0.05 mm.

▪ **TABLE 6.3**
Inside Diameter Measurements (mm) for Automobile Engine Piston Rings

Sample Number	Observations					$\bar{x}_i$	s_i
1	74.030	74.002	74.019	73.992	74.008	74.010	0.0148
2	73.995	73.992	74.001	74.011	74.004	74.001	0.0075
3	73.988	74.024	74.021	74.005	74.002	74.008	0.0147
4	74.002	73.996	73.993	74.015	74.009	74.003	0.0091
5	73.992	74.007	74.015	73.989	74.014	74.003	0.0122
6	74.009	73.994	73.997	73.985	73.993	73.996	0.0087
7	73.995	74.006	73.994	74.000	74.005	74.000	0.0055
8	73.985	74.003	73.993	74.015	73.988	73.997	0.0123
9	74.008	73.995	74.009	74.005	74.004	74.004	0.0055
10	73.998	74.000	73.990	74.007	73.995	73.998	0.0063
11	73.994	73.998	73.994	73.995	73.990	73.994	0.0029
12	74.004	74.000	74.007	74.000	73.996	74.001	0.0042
13	73.983	74.002	73.998	73.997	74.012	73.998	0.0105
14	74.006	73.967	73.994	74.000	73.984	73.990	0.0153
15	74.012	74.014	73.998	73.999	74.007	74.006	0.0073
16	74.000	73.984	74.005	73.998	73.996	73.997	0.0078
17	73.994	74.012	73.986	74.005	74.007	74.001	0.0106
18	74.006	74.010	74.018	74.003	74.000	74.007	0.0070
19	73.984	74.002	74.003	74.005	73.997	73.998	0.0085
20	74.000	74.010	74.013	74.020	74.003	74.009	0.0080
21	73.982	74.001	74.015	74.005	73.996	74.000	0.0122
22	74.004	73.999	73.990	74.006	74.009	74.002	0.0074
23	74.010	73.989	73.990	74.009	74.014	74.002	0.0119
24	74.015	74.008	73.993	74.000	74.010	74.005	0.0087
25	73.982	73.984	73.995	74.017	74.013	73.998	0.0162
					$\Sigma = 1,850.028$		0.2351
					$\bar{\bar{x}} = \quad 74.001$		$\bar{s} = 0.0094$

6.15. continued

(a) Set up $\bar{x}$ and R control charts on this process. Is the process in statistical control?

MTB > Stat > Control Charts > Variables Charts for Subgroups > Xbar-R

The process is in statistical control with no out-of-control signals, runs, trends, or cycles.

(b) Note that the control limits on the $\bar{x}$ chart in part (a) are identical to the control limits on the $\bar{x}$ chart in Example 6.3, where the limits were based on s. Will this always happen?

The control limits on the $\bar{x}$ charts in Example 6.3 were calculated using $\bar{S}$ to estimate σ, in this exercise $\bar{R}$ was used to estimate σ. They will not always be the same, and in general, the $\bar{x}$ control limits based on $\bar{S}$ will be slightly different than limits based on $\bar{R}$.

6.15. continued

(c) Estimate process capability for the piston-ring process. Estimate the percentage of piston rings produced that will be outside of the specification.

$$\hat{\sigma}_x = \overline{R} / d_2 = 0.02324 / 2.326 = 0.009991$$

$$\hat{C}_p = \frac{USL - LSL}{6\hat{\sigma}_x} = \frac{74.05 - 73.95}{6(0.009991)} = 1.668$$

The process is capable of meeting specifications.

MTB > Stat > Quality Tools > Capability Analysis > Normal
Under "Estimate" select Rbar as method to estimate standard deviation.

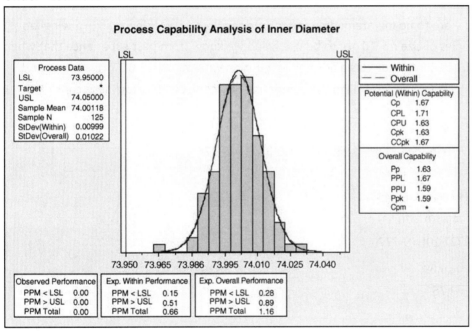

6.15.(c) continued

$$\hat{p} = \Pr\{x < LSL\} + \Pr\{x > USL\}$$

$$= \Pr\{x < 73.95\} + \Pr\{x > 74.05\}$$

$$= \Pr\{x < 73.95\} + \left[1 - \Pr\{x < 74.05\}\right]$$

$$= \Phi\left(\frac{73.95 - 74.00118}{0.009991}\right) + \left[1 - \Phi\left(\frac{74.05 - 74.00118}{0.009991}\right)\right]$$

$$= \Phi(-5.123) + 1 - \Phi(4.886)$$

$$= 0 + 1 - 1$$

$$= 0$$

6.17.

Control charts on $\bar{x}$ and s are to be maintained on the torque readings of a earing used in a wingflap actuator assembly. Samples of size $n = 10$ are to be used, and we know from past experience that when the process is in control, bearing torque has a normal distribution with mean $\mu = 80$ inch-pounds and standard deviation of $\sigma = 10$ inch pounds. Find the center line and control limits for these control charts.

$n = 10$; $\mu = 80$ in-lb; $\sigma_x = 10$ in-lb; and $A = 0.949$; $B_6 = 1.669$; $B_5 = 0.276$

$\text{centerline}_{\bar{x}} = \mu = 80$

$\text{UCL}_{\bar{x}} = \mu + A\sigma_x = 80 + 0.949(10) = 89.49$

$\text{LCL}_{\bar{x}} = \mu - A\sigma_x = 80 - 0.949(10) = 70.51$

$\text{centerline}_S = c_4\sigma_x = 0.9727(10) = 9.727$

$\text{UCL}_S = B_6\sigma_x = 1.669(10) = 16.69$

$\text{LCL}_S = B_5\sigma_x = 0.276(10) = 2.76$

6.19.

Samples of $n = 4$ items are taken from a process at regular intervals. A normally distributed quality characteristic is measured and $\bar{x}$ and s values are calculated for each sample. After 50 subgroups have been analyzed, we have

$$\sum_{i=1}^{50} \bar{x}_i = 1000; \ \sum_{i=1}^{50} S_i = 72; \ m = 50 \text{ subgroups}$$

(a) Compute the control limit for the $\bar{x}$ and s control charts

$$\bar{\bar{x}} = \frac{\sum_{i=1}^{50} \bar{x}_i}{m} = \frac{1000}{50} = 20; \ \bar{S} = \frac{\sum_{i=1}^{50} S_i}{m} = \frac{72}{50} = 1.44$$

$$UCL_{\bar{x}} = \bar{\bar{x}} + A_3\bar{S} = 20 + 1.628(1.44) = 22.34$$

$$LCL_{\bar{x}} = \bar{\bar{x}} - A_3\bar{S} = 20 - 1.628(1.44) = 17.66$$

$$UCL_S = B_4\bar{S} = 2.266(1.44) = 3.26$$

$$LCL_S = B_3\bar{S} = 0(1.44) = 0$$

(b) Assume that all points on both charts plot within the control limits. What are the natural tolerance limits of the process?

natural process tolerance limits: $\bar{\bar{x}} \pm 3\hat{\sigma}_x = \bar{\bar{x}} \pm 3\left(\dfrac{\bar{S}}{c_4}\right) = 20 \pm 3\left(\dfrac{1.44}{0.9213}\right) = [15.3, 24.7]$

(c) If the specification limits are 19 ± 4.0, what are your conclusions regarding the ability of the process to produce items conforming to specifications?

$$\hat{C}_p = \frac{USL - LSL}{6\hat{\sigma}_x} = \frac{+4.0 - (-4.0)}{6(1.44/0.9213)} = 0.85$$

The process is not capable.

6.19. continued

(d) Assuming that if an item exceeds the upper specification limit it can be reworked, and if it is below the lower specification limit it must be scrapped, what percent scrap and rework is the process now producing?

$$\hat{p}_{rework} = Pr\{x > USL\} = 1 - Pr\{x \le USL\} = 1 - \Phi\left(\frac{23 - 20}{1.44/0.9213}\right) = 1 - \Phi(1.919) = 1 - 0.9725 = 0.0275 \text{ or } 2.75\%.$$

$$\hat{p}_{scrap} = Pr\{x < LSL\} = \Phi\left(\frac{15 - 20}{1.44/0.9213}\right) = \Phi(-3.199) = 0.00069, \text{ or } 0.069\%$$

Total = 2.88% + 0.069% = 2.949%

(e) If the process were centered at $\mu = 19.0$, what would be the effect on percent scrap and rework?

$$\hat{p}_{rework} = 1 - \Phi\left(\frac{23 - 19}{1.44/0.9213}\right) = 1 - \Phi(2.56) = 1 - 0.99477 = 0.00523, \text{ or } 0.523\%$$

$$\hat{p}_{scrap} = \Phi\left(\frac{15 - 19}{1.44/0.9213}\right) = \Phi(-2.56) = 0.00523, \text{ or } 0.523\%$$

Total = 0.523% + 0.523% = 1.046%

Centering the process would reduce rework, but increase scrap. A cost analysis is needed to make the final decision. An alternative would be to work to improve the process by reducing variability.

6.21.

Parts manufactured by an injection molding process are subjected to a compressive strength test. Twenty samples of five parts each are collected, and the compressive strengths (in psi) are shown in Table 6E.12.

■ TABLE 6E.12
Strength Data for Exercise 6.21

Sample Number	x_1	x_2	x_3	x_4	x_5	$\bar{x}$	R
1	83.0	81.2	78.7	75.7	77.0	79.1	7.3
2	88.6	78.3	78.8	71.0	84.2	80.2	17.6
3	85.7	75.8	84.3	75.2	81.0	80.4	10.4
4	80.8	74.4	82.5	74.1	75.7	77.5	8.4
5	83.4	78.4	82.6	78.2	78.9	80.3	5.2
6	75.3	79.9	87.3	89.7	81.8	82.8	14.5
7	74.5	78.0	80.8	73.4	79.7	77.3	7.4
8	79.2	84.4	81.5	86.0	74.5	81.1	11.4
9	80.5	86.2	76.2	64.1	80.2	81.4	9.9
10	75.7	75.2	71.1	82.1	74.3	75.7	10.9
11	80.0	81.5	78.4	73.8	78.1	78.4	7.7
12	80.6	81.8	79.3	73.8	81.7	79.4	8.0
13	82.7	81.3	79.1	82.0	79.5	80.9	3.6
14	79.2	74.9	78.6	77.7	75.3	77.1	4.3
15	85.5	82.1	82.8	73.4	71.7	79.1	13.8
16	78.8	79.6	80.2	79.1	80.8	79.7	2.0
17	82.1	78.2	75.5	78.2	82.1	79.2	6.6
18	84.5	76.9	83.5	81.2	79.2	81.1	7.6
19	79.0	77.8	81.2	84.4	81.6	80.8	6.6
20	84.5	73.1	78.6	78.7	80.6	79.1	11.4

6.21. continued

(a) Establish $\bar{x}$ and R control charts for compressive strength using these data. Is the process in statistical control?

MTB > Stat > Control Charts > Variables Charts for Subgroups > Xbar-R

Yes, the process is in control—though we should watch for a possible cyclic pattern in the averages.

(b) After establishing the control charts in part (a), 15 new subgroups were collected and the compressive strengths are shown in Table 6E.13. Plot the $\bar{x}$ and R values against the control units from part (a) and draw conclusions.

■ **TABLE 6E.13**
New Data for Exercise 6.21, part (b)

Sample Number	x_1	x_2	x_3	x_4	x_5	$\bar{x}$	R
1	68.9	81.5	78.2	80.8	81.5	78.2	12.6
2	69.8	68.6	80.4	84.3	83.9	77.4	15.7
3	78.5	85.2	78.4	80.3	81.7	80.8	6.8
4	76.9	86.1	86.9	94.4	83.9	85.6	17.5
5	93.6	81.6	87.8	79.6	71.0	82.7	22.5
6	65.5	86.8	72.4	82.6	71.4	75.9	21.3
7	78.1	65.7	83.7	93.7	93.4	82.9	27.9
8	74.9	72.6	81.6	87.2	72.7	77.8	14.6
9	78.1	77.1	67.0	75.7	76.8	74.9	11.0
10	78.7	85.4	77.7	90.7	76.7	81.9	14.0
11	85.0	60.2	68.5	71.1	82.4	73.4	24.9
12	86.4	79.2	79.8	86.0	75.4	81.3	10.9
13	78.5	99.0	78.3	71.4	81.8	81.7	27.6
14	68.8	62.0	82.0	77.5	76.1	73.3	19.9
15	83.0	83.7	73.1	82.2	95.3	83.5	22.2

6.21.(b) continued

MTB > Stat > Control Charts > Variables Charts for Subgroups > Xbar-R

Under "Options, Estimate" use subgroups 1:20 to calculate control limits.

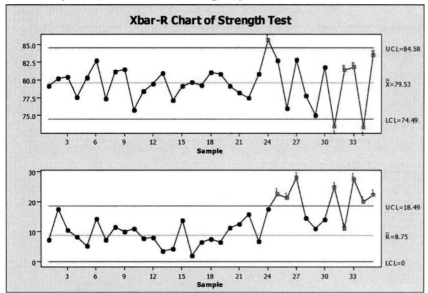

Test Results for R Chart of Ex6.15bSt
```
TEST 1. One point more than 3.00 standard deviations from center line.
Test Failed at points:   25, 26, 27, 31, 33, 34, 35
TEST 2. 9 points in a row on same side of center line.
Test Failed at points:   32, 33, 34, 35
```

A strongly cyclic pattern in the averages is now evident, but more importantly, there are several out-of-control points on the range chart.

6.23.

Consider the $\bar{x}$ and R charts you established in Exercise 6.7 using $n = 5$.

$n_{\text{old}} = 5; \; \bar{\bar{x}}_{\text{old}} = 34.00; \; \bar{R}_{\text{old}} = 4.7$

(a) Suppose that you wished to continue charting this quality characteristics using $\bar{x}$ and R charts based on a sample size of $n = 3$. What limits would be used on the $\bar{x}$ and R charts?

for $n_{\text{new}} = 3$

$$UCL_{\bar{x}} = \bar{\bar{x}}_{\text{old}} + A_{2(\text{new})} \left[\frac{d_{2(\text{new})}}{d_{2(\text{old})}} \right] \bar{R}_{\text{old}} = 34 + 1.023 \left[\frac{1.693}{2.326} \right] (4.7) = 37.50$$

$$LCL_{\bar{x}} = \bar{\bar{x}}_{\text{old}} - A_{2(\text{new})} \left[\frac{d_{2(\text{new})}}{d_{2(\text{old})}} \right] \bar{R}_{\text{old}} = 34 - 1.023 \left[\frac{1.693}{2.326} \right] (4.7) = 30.50$$

$$UCL_R = D_{4(\text{new})} \left[\frac{d_{2(\text{new})}}{d_{2(\text{old})}} \right] \bar{R}_{\text{old}} = 2.574 \left[\frac{1.693}{2.326} \right] (4.7) = 8.81$$

$$CL_R = \bar{R}_{\text{new}} = \left[\frac{d_{2(\text{new})}}{d_{2(\text{old})}} \right] \bar{R}_{\text{old}} = \left[\frac{1.693}{2.326} \right] (4.7) = 3.42$$

$$LCL_R = D_{3(\text{new})} \left[\frac{d_{2(\text{new})}}{d_{2(\text{old})}} \right] \bar{R}_{\text{old}} = 0 \left[\frac{1.693}{2.326} \right] (4.7) = 0$$

(b) What would be the impact of the decision you made in part (a) on the ability of the $\bar{x}$ chart to detect a 2σ shift in the mean?

The $\bar{x}$ control limits for $n = 5$ are "tighter" (31.29, 36.72) than those for $n = 3$ (30.50, 37.50). This means a 2σ shift in the mean would be detected more quickly with a sample size of $n = 5$.

6.23. continued

(c) Suppose you wished to continue charting this quality characteristic using $\overline{x}$ and R charts based on a sample size of $n = 8$. What limits would be used on the $\overline{x}$ and R charts?

$$\text{UCL}_{\overline{x}} = \overline{\overline{x}}_{\text{old}} + A_{2(\text{new})} \left[\frac{d_{2(\text{new})}}{d_{2(\text{old})}} \right] \overline{R}_{\text{old}} = 34 + 0.373 \left[\frac{2.847}{2.326} \right] (4.7) = 36.15$$

$$\text{LCL}_{\overline{x}} = \overline{\overline{x}}_{\text{old}} - A_{2(\text{new})} \left[\frac{d_{2(\text{new})}}{d_{2(\text{old})}} \right] \overline{R}_{\text{old}} = 34 - 0.373 \left[\frac{2.847}{2.326} \right] (4.7) = 31.85$$

$$\text{UCL}_{R} = D_{4(\text{new})} \left[\frac{d_{2(\text{new})}}{d_{2(\text{old})}} \right] \overline{R}_{\text{old}} = 1.864 \left[\frac{2.847}{2.326} \right] (4.7) = 10.72$$

$$\text{CL}_{R} = \overline{R}_{\text{new}} = \left[\frac{d_{2(\text{new})}}{d_{2(\text{old})}} \right] \overline{R}_{\text{old}} = \left[\frac{2.847}{2.326} \right] (4.7) = 5.75$$

$$\text{LCL}_{R} = D_{3(\text{new})} \left[\frac{d_{2(\text{new})}}{d_{2(\text{old})}} \right] \overline{R}_{\text{old}} = 0.136 \left[\frac{2.847}{2.326} \right] (4.7) = 0.78$$

(d) What is the impact of using $n = 8$ on the ability of the $\overline{x}$ chart to detect a 2σ shift in the mean?

The $\overline{x}$ control limits for $n = 8$ are even "tighter" (31.85, 36.15), increasing the ability of the chart to quickly detect the 2σ shift in process mean.

6.25.

Control charts for $\bar{x}$ and R are maintained for an important quality characteristic. The sample size is $n = 7$; $\bar{x}$ and R are computed for each sample. After 35 samples we have found that

$$\sum_{i=1}^{35} \bar{x}_i = 7,805 \text{ and } \sum_{i=1}^{35} R_i = 1,200$$

(a) Set up $\bar{x}$ and R charts using these data.

$$\bar{\bar{x}} = \frac{\sum_{i=1}^{35} \bar{x}_i}{m} = \frac{7805}{35} = 223; \quad \bar{R} = \frac{\sum_{i=1}^{35} R_i}{m} = \frac{1200}{35} = 34.29$$

$$\text{UCL}_{\bar{x}} = \bar{\bar{x}} + A_2\bar{R} = 223 + 0.419(34.29) = 237.37$$

$$\text{LCL}_{\bar{x}} = \bar{\bar{x}} - A_2\bar{R} = 223 - 0.419(34.29) = 208.63$$

$$\text{UCL}_R = D_4\bar{R} = 1.924(34.29) = 65.97$$

$$\text{LCL}_R = D_3\bar{R} = 0.076(34.29) = 2.61$$

(b) Assuming that both charts exhibit control, estimate the process mean and standard deviation.

$$\hat{\mu} = \bar{\bar{x}} = 223; \quad \hat{\sigma}_x = \bar{R}/d_2 = 34.29/2.704 = 12.68$$

(c) If the quality characteristic is normally distributed and if the specifications are 220 ± 35, can the process meet the specifications? Estimate the fraction nonconforming.

$$\hat{C}_p = \frac{\text{USL} - \text{LSL}}{6\hat{\sigma}_x} = \frac{+35 - (-35)}{6(12.68)} = 0.92 \text{, the process is not capable of meeting specifications.}$$

$$\hat{p} = \Pr\{x > \text{USL}\} + \Pr\{x < \text{LSL}\} = 1 - \Pr\{x < \text{USL}\} + \Pr\{x < \text{LSL}\} = 1 - \Pr\{x \le 255\} + \Pr\{x \le 185\}$$

$$= 1 - \Phi\left(\frac{255 - 223}{12.68}\right) + \Phi\left(\frac{185 - 223}{12.68}\right) = 1 - \Phi(2.52) + \Phi(-3.00) = 1 - 0.99413 + 0.00135 = 0.0072$$

(d) Assuming the variance to remain constant, state where the process mean should be located to minimize the fraction nonconforming. What would be the value of the fraction nonconforming under these conditions?

The process mean should be located at the nominal dimension, 220, to minimize non-conforming units.

$$\hat{p} = 1 - \Phi\left(\frac{255 - 220}{12.68}\right) + \Phi\left(\frac{185 - 220}{12.68}\right) = 1 - \Phi(2.76) + \Phi(-2.76) = 1 - 0.99711 + 0.00289 = 0.00578$$

6.27.

Samples of size $n = 5$ are collected from a process every half hour. After 50 samples have been collected, we calculate $\bar{x} = 20.0$ and $s = 1.5$. Assume that both charts exhibit control and that the quality characteristic is normally distributed.

(a) Estimate the process standard deviation.

$$\hat{\sigma}_x = \bar{S}/c_4 = 1.5/0.9400 = 1.60$$

(b) Find the control limits on the charts.

$$UCL_{\bar{x}} = \bar{\bar{x}} + A_3\bar{S} = 20.0 + 1.427(1.5) = 22.14$$
$$LCL_{\bar{x}} = \bar{\bar{x}} - A_3\bar{S} = 20.0 - 1.427(1.5) = 17.86$$
$$UCL_S = B_4\bar{S} = 2.089(1.5) = 3.13$$
$$LCL_S = B_3\bar{S} = 0(1.5) = 0$$

(c) If the process mean shifts to 22, what is the probability of concluding that the process is still in control?

$$Pr\{in\ control\} = Pr\{LCL \leq \bar{x} \leq UCL\} = Pr\{\bar{x} \leq UCL\} - Pr\{\bar{x} \leq LCL\}$$

$$= \Phi\left(\frac{22.14 - 22}{1.6/\sqrt{5}}\right) - \Phi\left(\frac{17.86 - 22}{1.6/\sqrt{5}}\right) = \Phi(0.20) - \Phi(-5.79)$$

$$= 0.57926 - 0 = 0.57926$$

6.31.

A critical dimension of a machined part has specifications 100 ± 10. Control chart analysis indicates tht the process is in control with $\bar{x} = 104$ and $\bar{R} = 9.30$. The control charts use samples of size $n = 5$. If we assume that the characteristic is normally distributed, can the mean be located (by adjusting the tool position) so that all output meets specifications?

$\hat{\sigma}_x = \bar{R}/d_2 = 9.30/2.326 = 3.998$ and $6\hat{\sigma}_x = 6(3.998) = 23.99$ is larger than the width of the tolerance band, $2(10) = 20$. So, even if the mean is located at the nominal dimension, 100, not all of the output will meet specification.

$$\hat{C}_p = \frac{USL - LSL}{6\hat{\sigma}_x} = \frac{+10 - (-10)}{6(3.998)} = 0.8338$$

6.33.

Samples of $n=5$ units are taken from a process every hour. The $\bar{x}$ and R values for a particular quality characteristic are determined. After 25 sample have been collected, we calculate $\bar{\bar{x}}=20$ and $\bar{R}=4.56$.

(a) What are the three-sigma control limits for the $\bar{x}$ and R ?

$UCL_{\bar{x}} = \bar{\bar{x}} + A_2\bar{R} = 20 + 0.577(4.56) = 22.63$
$LCL_{\bar{x}} = \bar{\bar{x}} - A_2\bar{R} = 20 - 0.577(4.56) = 17.37$
$UCL_R = D_4\bar{R} = 2.114(4.56) = 9.64$
$LCL_R = D_3\bar{R} = 0(4.56) = 0$

(b) Both charts exhibit control. Estimate the process standard deviation.

$\hat{\sigma}_x = \bar{R}/d_2 = 4.56/2.326 = 1.96$

(c) Assume that the process output is normally distributed. If he specifications are 19 ± 5, what are your conclusions regarding the process capability?

$\hat{C}_p = \dfrac{USL - LSL}{6\hat{\sigma}_x} = \dfrac{+5-(-5)}{6(1.96)} = 0.85$, so the process is not capable of meeting specifications.

(d) If the process mean shifts to 24, what is the probability of not detecting this shift on the first subsequent sample?

$Pr\{\text{not detect}\} = Pr\{LCL \le \bar{x} \le UCL\} = Pr\{\bar{x} \le UCL\} - Pr\{\bar{x} \le LCL\}$

$$= \Phi\left(\dfrac{UCL_{\bar{x}} - \mu_{new}}{\hat{\sigma}_x/\sqrt{n}}\right) - \Phi\left(\dfrac{LCL_{\bar{x}} - \mu_{new}}{\hat{\sigma}_x/\sqrt{n}}\right) = \Phi\left(\dfrac{22.63-24}{1.96/\sqrt{5}}\right) - \Phi\left(\dfrac{17.37-24}{1.96/\sqrt{5}}\right)$$

$$= \Phi(-1.56) + \Phi(-7.56) = 0.05938 - 0 = 0.05938$$

6.35.

Continuation of Exercise 6.34. Table 6E.15 contains 10 new subgroups of thickness data. Plot this data on the control charts constructed in Exercise 6.26(a). Is the process in statistical control?

■ TABLE 6E.15
Additional Thickness Data for
Exercise 6.35.

Subgroup	x_1	x_2	x_3	x_4
21	454	449	443	461
22	449	441	444	455
23	442	442	442	450
24	443	452	438	430
25	446	459	457	457
26	454	448	445	462
27	458	449	453	438
28	450	449	445	451
29	443	440	443	451
30	457	450	452	437

MTB > Stat > Control Charts > Variables Charts for Subgroups > Xbar-R

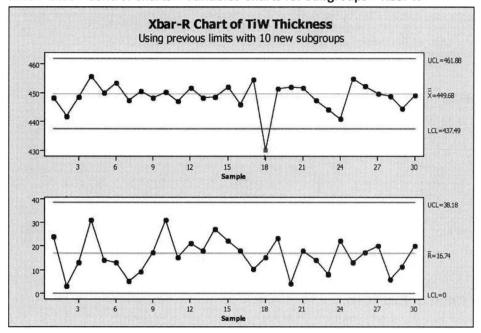

6.35. continued

Test Results for Xbar Chart of Ex6.27Th
```
TEST 1. One point more than 3.00 standard deviations from center line.
Test Failed at points:  18
```

The process continues to be in a state of statistical control.

6.37.

Rework Exercises 6.34 and 6.35 using $\bar{x}$ and s control charts.

The process is out of control on the $\bar{x}$ chart at subgroup 18. After finding assignable cause, exclude subgroup 18 from control limits calculations:

MTB > Stat > Control Charts > Variables Charts for Subgroups > Xbar-S

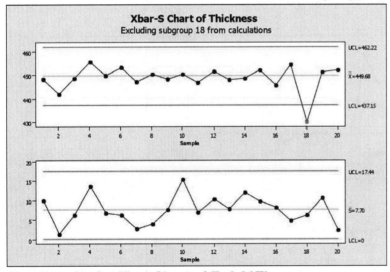

Test Results for Xbar Chart of Ex6.26Th
```
TEST 1. One point more than 3.00 standard deviations from center line.
Test Failed at points:  18
```

No additional subgroups are beyond the control limits, so these limits can be used for future production.

6.37. continued

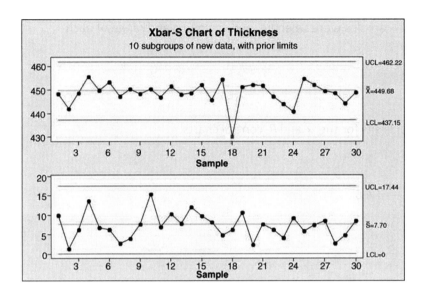

The process remains in statistical control.

6.39.

An $\bar{x}$ chart has a center line of 100, uses three-sigma control limits, and is based on a sample size of four. The process standard deviation is known to be six. If the process mean shifts from 100 to 92, what is the probability of detecting this shift on the first sample following the shift?

$\mu_0 = 100$; $\quad L = 3$; $\quad n = 4$; $\quad \sigma = 6$; $\quad \mu_1 = 92$

$k = \left(\mu_1 - \mu_0\right)/\sigma = \left(92 - 100\right)/6 = -1.33$

Pr{detecting shift on 1st sample} $= 1 - \text{Pr\{not detecting shift on 1st sample\}}$

$$= 1 - \beta$$
$$= 1 - \left[\Phi\left(L - k\sqrt{n}\right) - \Phi\left(-L - k\sqrt{n}\right) \right]$$
$$= 1 - \left[\Phi\left(3 - (-1.33)\sqrt{4}\right) - \Phi\left(-3 - (-1.33)\sqrt{4}\right) \right]$$
$$= 1 - \left[\Phi(5.66) - \Phi(-0.34) \right]$$
$$= 1 - \left[1 - 0.37 \right]$$
$$= 0.37$$

6.41.

Control charts on $\bar{x}$ and R for samples of size $n-5$ are to be maintained on the tensile strength in pounds of a yarn. To start the charts, 30 samples were selected, and the mean and range of each computed. This yields

$$\sum_{i=1}^{30} \bar{x}_i = 607.8 \text{ and } \sum_{i=1}^{30} R_i = 144$$

(a) Compute the center line and control limits for the $\bar{x}$ and R control charts.

$$\bar{\bar{x}} = \frac{\sum_{i=1}^{m} \bar{x}_i}{m} = \frac{607.8}{30} = 20.26; \quad \bar{R} = \frac{\sum_{i=1}^{m} R_i}{m} = \frac{144}{30} = 4.8$$

$$\text{UCL}_{\bar{x}} = \bar{\bar{x}} + A_2\bar{R} = 20.26 + 0.577(4.8) = 23.03$$

$$\text{LCL}_{\bar{x}} = \bar{\bar{x}} - A_2\bar{R} = 20.26 - 0.577(4.8) = 17.49$$

$$\text{UCL}_R = D_4\bar{R} = 2.114(4.8) = 10.147$$

$$\text{LCL}_R = D_3\bar{R} = 0(4.8) = 0$$

(b) Suppose both charts exhibit control. There is a single lower specification limit of 16 lb. If strength is normally distributed, what fraction of yarn would fail to meet specifications?

$$\hat{\sigma}_x = \bar{R}/d_2 = 4.8/2.326 = 2.064$$

$$\hat{p} = \Pr\{x < \text{LSL}\} = \Phi\left(\frac{16 - 20.26}{2.064}\right) = \Phi(-2.064) = 0.0195$$

6.43.

Continuation of Exercise 6.42. Reconsider the data from Exercise 6.42 and establish $\bar{x}$ and R charts with appropriate trial control limits. Revise these trial limits as necessary to produce a set of control charts for monitoring future production. Suppose that the new data in Table 6E.19 are observed.

■ **TABLE 6E.19**
New Data for Exercise 6.43

Sample Number	x_1	x_2	x_3	x_4	x_5
16	2	10	9	6	5
17	1	9	5	9	4
18	0	9	8	2	5
19	-3	0	5	1	4
20	2	10	9	3	1
21	-5	4	0	6	-1
22	0	2	-5	4	6
23	10	0	3	1	5
24	-1	2	5	6	-3
25	0	-1	2	5	-2

(a) Plot these new observations on the control chart. What conclusions can you draw about process stability?

MTB > Stat > Control Charts > Variables Charts for Subgroups > R
Under "Options, Estimate" use subgroups 1:11 and 13:15, and select Rbar.

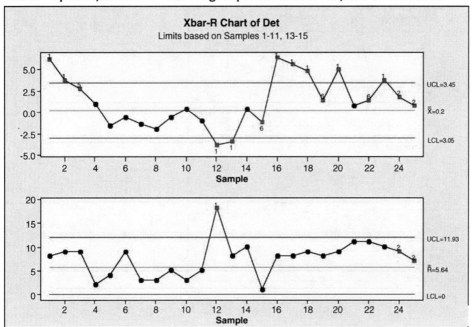

6.43.(a) continued

Test Results for Xbar Chart of Ex6.35Det

TEST 1. One point more than 3.00 standard deviations from center line.
Test Failed at points: 1, 2, 12, 13, 16, 17, 18, 20, 23
TEST 2. 9 points in a row on same side of center line.
Test Failed at points: 24, 25
TEST 5. 2 out of 3 points more than 2 standard deviations from center line (on
    one side of CL).
Test Failed at points: 2, 3, 13, 17, 18, 20
TEST 6. 4 out of 5 points more than 1 standard deviation from center line (on
    one side of CL).
Test Failed at points: 15, 19, 20, 22, 23, 24

Test Results for R Chart of Ex6.35Det

TEST 1. One point more than 3.00 standard deviations from center line.
Test Failed at points: 12
TEST 2. 9 points in a row on same side of center line.
Test Failed at points: 24, 25

We are trying to establish trial control limits from the first 15 samples to monitor future production. Note that samples 1, 2, 12, and 13 are out of control on the $\bar{x}$ chart. If these samples are removed and the limits recalculated, sample 3 is also out of control on the $\bar{x}$ chart. Removing sample 3 gives

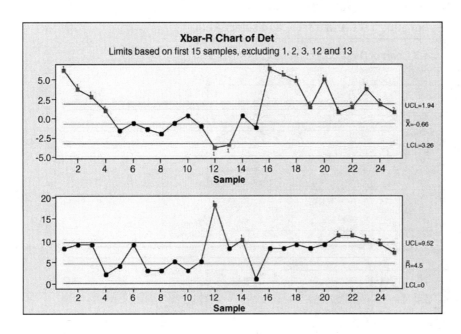

6.43.(a) continued

Sample 14 is now out of control on the R chart. No additional samples are out of control on the $\bar{x}$ chart. While the limits on the above charts may be used to monitor future production, the fact that 6 of 15 samples were out of control and eliminated from calculations is an early indication of process instability.

Given the large number of points after sample 15 beyond both the $\bar{x}$ and R control limits on the charts above, the process appears to be unstable.

(b) Use all 25 observations to revise the control limits for the $\bar{x}$ and R charts. What conclusions can you draw now about the process?

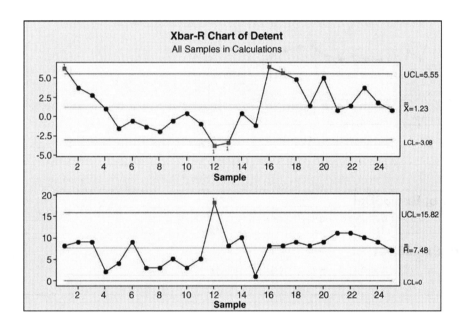

With Test 1 only:

Test Results for Xbar Chart of Ex6.35Det
TEST 1. One point more than 3.00 standard deviations from center line.
Test Failed at points: 1, 12, 13, 16, 17
Test Results for R Chart of Ex6.35Det
TEST 1. One point more than 3.00 standard deviations from center line.
Test Failed at points: 12

6.43.(b) continued

Removing samples 1, 12, 13, 16, and 17 from calculations:

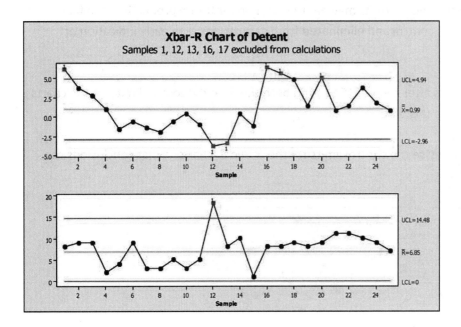

With Test 1 only:

Test Results for Xbar Chart of Ex6.35Det
TEST 1. One point more than 3.00 standard deviations from center line.
Test Failed at points: 1, 12, 13, 16, 17, 20
Test Results for R Chart of Ex6.35Det
TEST 1. One point more than 3.00 standard deviations from center line.
Test Failed at points: 12

6.43.(b) continued

Sample 20 is now also out of control. Removing sample 20 from calculations,

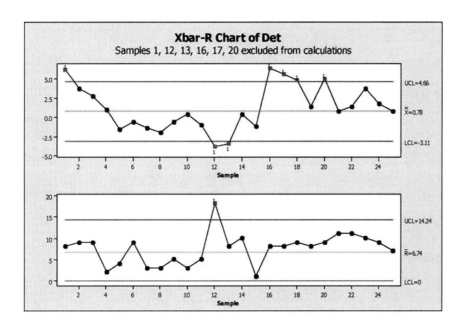

With Test 1 only:

Test Results for Xbar Chart of Ex6.35Det
```
TEST 1. One point more than 3.00 standard deviations from center line.
Test Failed at points:  1, 12, 13, 16, 17, 18, 20
```
Test Results for R Chart of Ex6.35Det
```
TEST 1. One point more than 3.00 standard deviations from center line.
Test Failed at points:  12
```

Sample 18 is now out-of-control, for a total 7 of the 25 samples, with runs of points both above and below the centerline. This suggests that the process is inherently unstable, and that the sources of variation need to be identified and removed.

6.45.

Control charts for $\bar{x}$ and R are maintained on the tensile strength of a metal fastener. Atfer 30 samples of size $n = 6$ are analyzed, we find that

$$\sum_{i=1}^{30} \bar{x}_i = 12{,}870 \text{ and } \sum_{i=1}^{30} R_i = 1350$$

(a) Compute control limits on the R chart.

$$\bar{R} = \frac{\sum_{i=1}^{m} R_i}{m} = \frac{1350}{30} = 45.0$$

$$\text{UCL}_R = D_4 \bar{R} = 2.004(45.0) = 90.18$$

$$\text{LCL}_R = D_3 \bar{R} = 0(45.0) = 0$$

(b) Assuming that the R chart exhibits control, estimate the parameters μ and σ.

$$\hat{\mu} = \bar{\bar{x}} = \frac{\sum_{i=1}^{m} \bar{x}_i}{m} = \frac{12{,}870}{30} = 429.0$$

$$\hat{\sigma}_x = \bar{R} / d_2 = 45.0 / 2.534 = 17.758$$

(c) If the process output is normally distributed, and if the specifications are 440 ± 40, can the process meet the specifications? Estimate the fraction nonconforming.

USL = 440 + 40 = 480; LSL = 440 - 40 = 400

$$\hat{C}_p \frac{\text{USL} - \text{LSL}}{6\hat{\sigma}_x} = \frac{480 - 400}{6(17.758)} = 0.751$$

$$\hat{p} = 1 - \Phi\left(\frac{480 - 429}{17.758}\right) + \Phi\left(\frac{400 - 429}{17.758}\right) = 1 - \Phi(2.87) + \Phi(-1.63) = 1 - 0.9979 + 0.0516 = 0.0537$$

(d) If the variance remains constant, where should the mean be located to minimize the fraction nonconforming?

To minimize fraction nonconforming the mean should be located at the nominal dimension (440) for a constant variance.

6.47.

An $\bar{x}$ chart on a normally distributed quality characteristic is to be established with the standard values $\mu = 100$, $\sigma = 8$, and $n = 4$. Find the following:

(a) The two-sigma control limits.

$n = 4; \quad \mu = 100; \quad \sigma_x = 8$

$UCL_{\bar{x}} = \mu + 2\sigma_{\bar{x}} = \mu + 2\left(\sigma_x / \sqrt{n}\right) = 100 + 2\left(8/\sqrt{4}\right) = 108$

$LCL_{\bar{x}} = \mu - 2\sigma_{\bar{x}} = \mu - 2\left(\sigma_x / \sqrt{n}\right) = 100 - 2\left(8/\sqrt{4}\right) = 92$

(b) The 0.005 probability limits.

$k = Z_{\alpha/2} = Z_{0.005/2} = Z_{0.0025} = 2.807$

$UCL_{\bar{x}} = \mu + k\sigma_{\bar{x}} = \mu + k\left(\sigma_x / \sqrt{n}\right) = 100 + 2.807\left(8/\sqrt{4}\right) = 111.228$

$LCL_{\bar{x}} = \mu - k\sigma_{\bar{x}} = \mu - k\left(\sigma_x / \sqrt{n}\right) = 100 - 2.807\left(8/\sqrt{4}\right) = 88.772$

6.49.

Consider the $\bar{x}$ chart defined in Exercise 6.48. Find the ARL_1 for the chart.

$$ARL_1 = \frac{1}{1-\beta} = \frac{1}{1-\text{Pr\{not detect\}}} = \frac{1}{1-0.6658} = 2.992$$

6.51.

Statistical monitoring of a quality characteristic uses both an $\bar{x}$ and s chart. The charts are to be based on the standard values $\mu = 200$ and $\sigma_x = 10$, with $n = 4$.

(a) Find three-sigma control limits for the s chart.

$\text{centerline}_s = c_4\sigma = 0.9213(10) = 9.213$

$UCL_s = B_6\sigma_x = 2.088(10) = 20.88$

$LCL_s = B_5\sigma_x = 0(10) = 0$

6.51. continued

(b) Find a center line and control limits for the $\bar{x}$ chart such that the probability of a type I error is 0.05.

$$k = Z_{\alpha/2} = Z_{0.05/2} = Z_{0.025} = 1.96$$

$$\text{UCL}_{\bar{x}} = \mu + k\sigma_{\bar{x}} = \mu + k\left(\sigma_x/\sqrt{n}\right) = 200 + 1.96\left(10/\sqrt{4}\right) = 209.8$$

$$\text{LCL}_{\bar{x}} = \mu - k\sigma_{\bar{x}} = \mu - k\left(\sigma_x/\sqrt{n}\right) = 200 - 1.96\left(10/\sqrt{4}\right) = 190.2$$

6.53.

Thirty samples each of size 7 have been collected to establish control over a process. The following data were collected:

$$\sum_{i=1}^{30} \bar{x}_i = 2700 \text{ and } \sum_{i=1}^{30} R_i = 120$$

(a) Calculate trial control limits for the two charts.

$$\bar{\bar{x}} = \frac{\sum_{i=1}^{m}\bar{x}_i}{m} = \frac{2700}{30} = 90; \quad \bar{R} = \frac{\sum_{i=1}^{m}R_i}{m} = \frac{120}{30} = 4$$

$$\text{UCL}_{\bar{x}} = \bar{\bar{x}} + A_2\bar{R} = 90 + 0.419(4) = 91.676$$

$$\text{LCL}_{\bar{x}} = \bar{\bar{x}} - A_2\bar{R} = 90 - 0.419(4) = 88.324$$

$$\text{UCL}_R = D_4\bar{R} = 1.924(4) = 7.696$$

$$\text{LCL}_R = D_3\bar{R} = 0.076(4) = 0.304$$

(b) On the assumption that the R chart is in control, estimate the process standard deviation.

$$\hat{\sigma}_x = \bar{R}/d_2 = 4/2.704 = 1.479$$

6.53. continued

(c) Suppose an s chart were desired. What would be the appropriate control limits and center line?

$$\bar{S} = c_4\hat{\sigma}_x = 0.9594(1.479) = 1.419$$
$$UCL_s = 1.882(1.419) = 2.671$$
$$LCL_s = 0.118(1.419) = 0.167$$

6.55.

$\bar{x}$ and R charts with $n = 4$ are used to monitor a normally distributed quality characteristic. The control chart parameters are

$\bar{x}$ Chart	R Chart
UCL = 815	UCL = 46.98
Center line = 800	Center line = 20.59
LCL = 785	LCL = 0

Both charts exhibit control. What is the probability that a shift in the process mean to 790 will be detected on the first sample following the shift?

$$\hat{\sigma}_x = \bar{R}/d_2 = 20.59/2.059 = 10$$

$$Pr\{\text{detect shift on 1st sample}\} = Pr\{\bar{x} < LCL\} + Pr\{\bar{x} > UCL\} = Pr\{\bar{x} < LCL\} + 1 - Pr\{\bar{x} \le UCL\}$$

$$= \Phi\left(\frac{LCL - \mu_{new}}{\sigma_{\bar{x}}}\right) + 1 - \Phi\left(\frac{UCL - \mu_{new}}{\sigma_{\bar{x}}}\right) = \Phi\left(\frac{785 - 790}{10/\sqrt{4}}\right) + 1 - \Phi\left(\frac{815 - 790}{10/\sqrt{4}}\right)$$

$$= \Phi(-1) + 1 - \Phi(5) = 0.1587 + 1 - 1.0000 = 0.1587$$

6.57.

Control charts for $\bar{x}$ and R are in use with the following parameters:

$\bar{x}$ Chart	R Chart
UCL = 363.0	UCL = 16.18
Center line = 360.0	Center line = 8.91
LCL = 357.0	LCL = 1.64

The sample size is $n = 9$. Both charts exhibit control. The quality characteristic is normally distributed.

(a) What is the α-risk associated with the $\bar{x}$ chart?

$$\hat{\sigma}_x = \bar{R}/d_2 = 8.91/2.970 = 3.000$$

$$\alpha = \Pr\{\bar{x} < LCL\} + \Pr\{\bar{x} > UCL\} = \Phi\left(\frac{LCL - \bar{\bar{x}}}{\sigma_{\bar{x}}}\right) + 1 - \Phi\left(\frac{UCL - \bar{\bar{x}}}{\sigma_{\bar{x}}}\right) = \Phi\left(\frac{357 - 360}{3/\sqrt{9}}\right) + 1 - \Phi\left(\frac{363 - 360}{3/\sqrt{9}}\right)$$

$$= \Phi(-3) + 1 - \Phi(3) = 0.0013 + 1 - 0.9987 = 0.0026$$

(b) Specifications on this quality characteristic are 358 ± 6. What are your conclusions regarding the ability of the process to produce items within specifications?

$$\hat{C}_p = \frac{USL - LSL}{6\hat{\sigma}_x} = \frac{+6 - (-6)}{6(3)} = 0.667$$

The process is not capable of producing all items within specification.

(c) Suppose the mean shifts to 357. What is the probability that the shift will not be detected on the first sample following the shift?

$$\mu_{new} = 357$$

$$\Pr\{\text{not detect on 1st sample}\} = \Pr\{LCL \leq \bar{x} \leq UCL\} = \Phi\left(\frac{UCL - \mu_{new}}{\hat{\sigma}_x/\sqrt{n}}\right) - \Phi\left(\frac{LCL - \mu_{new}}{\hat{\sigma}_x/\sqrt{n}}\right)$$

$$= \Phi\left(\frac{363 - 357}{3/\sqrt{9}}\right) - \Phi\left(\frac{357 - 357}{3/\sqrt{9}}\right) = \Phi(6) - \Phi(0) = 1.0000 - 0.5000 = 0.5000$$

6.57. continued

(d) What would be the appropriate control limits for the $\bar{x}$ chart if the type I error probability were to be 0.01?

$$\alpha = 0.01; \quad k = Z_{\alpha/2} = Z_{0.01/2} = Z_{0.005} = 2.576$$

$$\text{UCL}_{\bar{x}} = \bar{\bar{x}} + k\sigma_{\bar{x}} = \bar{\bar{x}} + k\left(\hat{\sigma}_x/\sqrt{n}\right) = 360 + 2.576\left(3/\sqrt{9}\right) = 362.576$$

$$\text{LCL}_{\bar{x}} = 360 - 2.576\left(3/\sqrt{9}\right) = 357.424$$

6.59.

Control charts for $\bar{x}$ and s have been maintained on a process and have exhibited statistical control. The sample size is $n = 6$. The control chart parameters are as follows:

$\bar{x}$ Chart	s Chart
UCL = 708.20	UCL = 3.420
Center line = 706.00	Center line = 1.738
LCL = 703.80	LCL = 0.052

(a) Estimate the mean and standard deviation of the process.

$$\hat{\mu} = \bar{\bar{x}} = 706.00; \quad \hat{\sigma}_x = \bar{s}/c_4 = 1.738/0.9515 = 1.827$$

(b) Estimate the natural tolerance limits for the process.

$$\text{UNTL} = \bar{\bar{x}} + 3\hat{\sigma}_x = 706 + 3(1.827) = 711.48$$

$$\text{LNTL} = 706 - 3(1.827) = 700.52$$

(c) Assume that the process output is well modeled by a normal distribution. If specifications are 703 and 709, estimate the fraction nonconforming.

$$\hat{p} = \Pr\{x < \text{LSL}\} + \Pr\{x > \text{USL}\} = \Phi\left(\frac{\text{LSL} - \bar{\bar{x}}}{\hat{\sigma}_x}\right) + 1 - \Phi\left(\frac{\text{USL} - \bar{\bar{x}}}{\hat{\sigma}_x}\right) = \Phi\left(\frac{703 - 706}{1.827}\right) + 1 - \Phi\left(\frac{709 - 706}{1.827}\right)$$

$$= \Phi(-1.642) + 1 - \Phi(1.642) = 0.0503 + 1 - 0.9497 = 0.1006$$

6.59. continued

(d) Suppose the process mean shifts to 702.00 while the standard deviation remains constant. What is the probability of an out-of-control signal occurring on the first sample following the shift?

$$\text{Pr\{detect on 1st sample\}} = \text{Pr}\{\bar{x} < \text{LCL}\} + \text{Pr}\{\bar{x} > \text{UCL}\}$$

$$= \Phi\left(\frac{\text{LCL} - \mu_{new}}{\sigma_{\bar{x}}}\right) + 1 - \Phi\left(\frac{\text{UCL} - \mu_{new}}{\sigma_{\bar{x}}}\right) = \Phi\left(\frac{703.8 - 702}{1.827/\sqrt{6}}\right) + 1 - \Phi\left(\frac{708.2 - 702}{1.827/\sqrt{6}}\right)$$

$$= \Phi(2.41) + 1 - \Phi(8.31) = 0.9920 + 1 - 1.0000 = 0.9920$$

(e) For the shift in part (d), what is the probability of detecting the shift by at least the third subsequent sample?

$$\text{Pr\{detect by 3rd sample\}} = 1 - \text{Pr\{not detect by 3rd sample\}}$$

$$= 1 - (\text{Pr\{not detect\}})^3 = 1 - (1 - 0.9920)^3 = 1.0000$$

6.61.

One-pound coffee cans are filled by a machine, sealed, and then weighted automatically. After adjusting for the weight of the can, any package that weighs less than 16 oz is cut out of the conveyor. The weights of 25 successive cans are shown in Table 6E.20. Set up a moving range control chart and a control chart for individuals. Estimate the mean and standard deviation of the amount of coffee packed in each can. Is it reasonable to assume that can weight is normally distributed? If the process remains in control at this level, what percentage of cans will be underfilled?

■ TABLE 6E.20
Can Weight Data for Exercise 6.61

Can Number	Weight	Can Number	Weight
1	16.11	14	16.12
2	16.08	15	16.10
3	16.12	16	16.08
4	16.10	17	16.13
5	16.10	18	16.15
6	16.11	19	16.12
7	16.12	20	16.10
8	16.09	21	16.08
9	16.12	22	16.07
10	16.10	23	16.11
11	16.09	24	16.13
12	16.07	25	16.10
13	16.13		

6.61. continued

MTB > Stat > Control Charts > Variables Charts for Individuals > I-MR

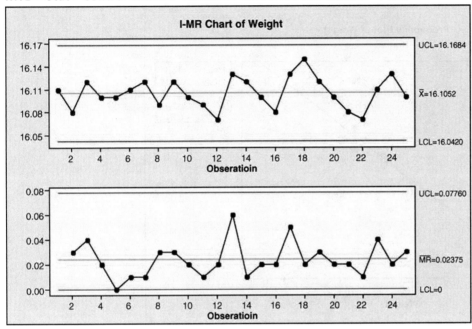

There may be a "sawtooth" pattern developing on the Individuals chart.

$$\bar{\bar{x}} = 16.1052; \quad \hat{\sigma}_x = 0.021055; \quad \overline{MR2} = 0.02375$$

6.61. continued

MTB > Stat > Basic Statistics > Normality Test

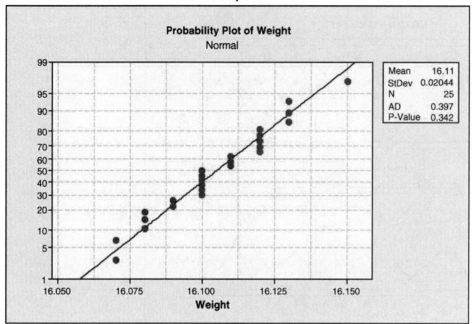

Visual examination of the normal probability indicates that the assumption of normally distributed coffee can weights is valid.

$$\%\text{underfilled} = 100\% \times \Pr\{x < 16 \text{ oz}\}$$

$$= 100\% \times \Phi\left(\frac{16 - 16.1052}{0.021055}\right) = 100\% \times \Phi(-4.9964) = 0.00003\%$$

6.63.

The viscosity of a polymer is measured hourly. Measurements for the last 20 hours are shown in table 6E.22.

■ **TABLE 6E.22**
Viscosity Data for Exercise 6.63

Test	Viscosity	Test	Viscosity
1	2838	11	3174
2	2785	12	3102
3	3058	13	2762
4	3064	14	2975
5	2996	15	2719
6	2882	16	2861
7	2878	17	2797
8	2920	18	3078
9	3050	19	2964
10	2870	20	2805

(a) Does viscosity follow a normal distribution?

MTB > Stat > Basic Statistics > Normality Test

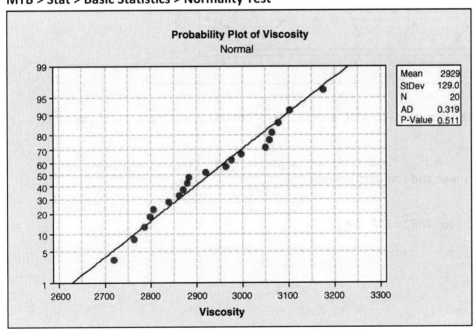

Viscosity measurements do appear to follow a normal distribution.

6.63. continued

(b) Set up a control chart on viscosity and a moving range chart. Does the process exhibit statistical control?

MTB > Stat > Control Charts > Variables Charts for Individuals > I-MR

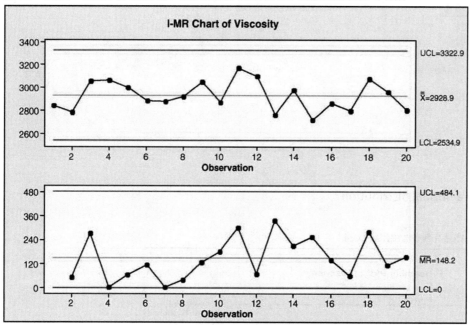

The process appears to be in statistical control, with no out-of-control points, runs, trends, or other patterns.

(c) Estimate the process mean and standard deviation.

$$\hat{\mu} = \bar{\bar{x}} = 2928.9; \quad \hat{\sigma}_x = 131.346; \quad \overline{MR2} = 148.158$$

6.65.

(a) Thirty observations on the oxide thickness of individual silicon wafers are shown in table 6E.23. Use these data to set up a control chart on oxide thickness and a moving range chart. Does the process exhibit statistical control? Does oxide thickness follow a normal distribution?

▦ **TABLE 6E.23**
Data. for Exercise 6.65

Wafer	Oxide Thickness	Wafer	Oxide Thickness
1	45.4	16	58.4
2	48.6	17	51.0
3	49.5	18	41.2
4	44.0	19	47.1
5	50.9	20	45.7
6	55.2	21	60.6
7	45.5	22	51.0
8	52.8	23	53.0
9	45.3	24	56.0
10	46.3	25	47.2
11	53.9	26	48.0
12	49.8	27	55.9
13	46.9	28	50.0
14	49.8	29	47.9
15	45.1	30	53.4

MTB > Stat > Control Charts > Variables Charts for Individuals > I-MR

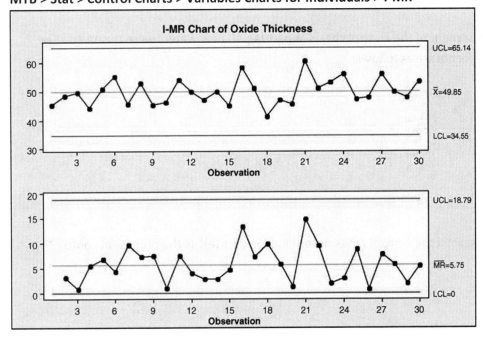

The process is in statistical control.

6.65.(a) continued

MTB > Stat > Basic Statistics > Normality Test

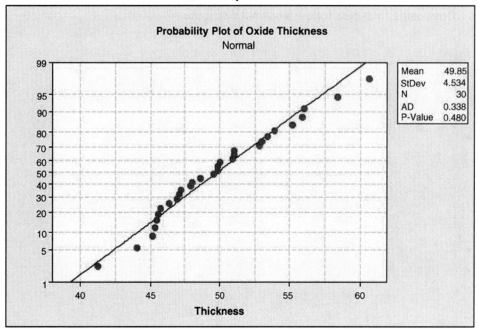

The normality assumption is reasonable.

(b) Following the establishment of the control charts in part (a), 10 new wafers were observed. The oxide thickness measurements are as follows:

Wafer	Oxide Thickness	Wafer	Oxide Thickness
1	54.3	6	51.5
2	57.5	7	58.4
3	64.8	8	67.5
4	62.1	9	61.1
5	59.6	10	63.3

Plot these observations against the control limits determined in part (a). Is the process in control?

6.65.(b) continued

MTB > Stat > Control Charts > Variables Charts for Individuals > I-MR

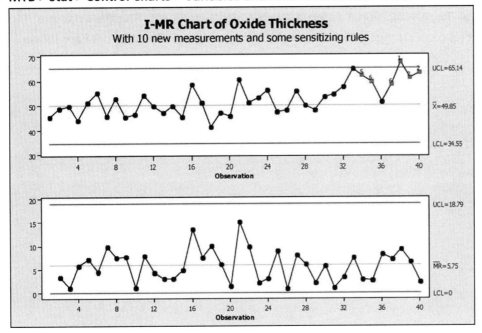

Test Results for I Chart of Ex6.57bTh

```
TEST 1. One point more than 3.00 standard deviations from center line.
Test Failed at points:   38
TEST 2. 9 points in a row on same side of center line.
Test Failed at points:   38, 39, 40
TEST 5. 2 out of 3 points more than 2 standard deviations from center line (on
        one side of CL).
Test Failed at points:   34, 39, 40
TEST 6. 4 out of 5 points more than 1 standard deviation from center line (on
        one side of CL).
Test Failed at points:   35, 37, 38, 39, 40
```

We have turned on some of the sensitizing rules in Minitab to illustrate their use. There is a run above the centerline, several 4 of 5 beyond 1 sigma, and several 2 of 3 beyond 2 sigma on the *x* chart. However, even without use of the sensitizing rules, it is clear that the process is out of control during this period of operation.

6.65. continued

(c) Suppose the assignable cause responsible for the out-of-control signal in part (b) is discovered and removed from the process. Twenty additional wafers are subsequently sampled. Plot the oxide thickness against the part (a) control limits. What conclusions can you draw? The new data are shown in Table 6E.25.

■ TABLE 6E.25
Additional Data for Exercise 6.65, part (c)

Wafer	Oxide Thickness	Wafer	Oxide Thickness
1	43.4	11	50.0
2	46.7	12	61.2
3	44.8	13	46.9
4	51.3	14	44.9
5	49.2	15	46.2
6	46.5	16	53.3
7	48.4	17	44.1
8	50.1	18	47.4
9	53.7	19	51.3
10	45.6	20	42.5

MTB > Stat > Control Charts > Variables Charts for Individuals > I-MR

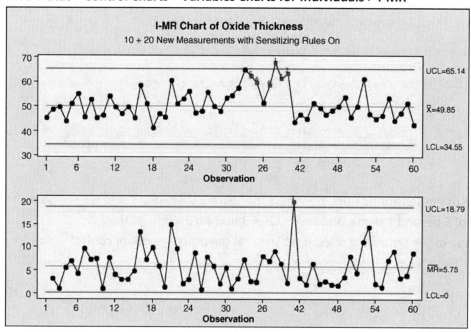

The process has been returned to a state of statistical control.

6.69.

In 1879, A.A. Michelson measured the velocity of light in air using a modification of a method proposed by the French physicist Foucault. Twenty of these measurements are in table 6E.27 (the value reported is in kilometers per second and has 299,000 subtracted from it). Use these data to set up individuals and moving range control charts. Is there some evidence that the measurements of the velocity of light are normally distributed? O the measurements exhibit statistical control? Revise the control limits if necessary.

■ **TABLE 6E.27**
Velocity of Light Data for Exercise 6.69

Measurement	Velocity	Measurement	Velocity
1	850	11	850
2	1000	12	810
3	740	13	950
4	980	14	1000
5	900	15	980
6	930	16	1000
7	1070	17	980
8	650	18	960
9	930	19	880
10	760	20	960

MTB > Stat > Basic Statistics > Normality Test

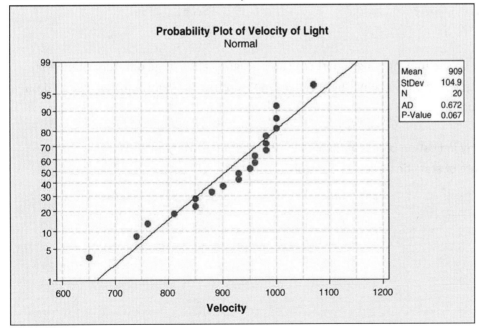

Velocity of light measurements are approximately normally distributed.

6.69. continued

MTB > Stat > Control Charts > Variables Charts for Individuals > I-MR

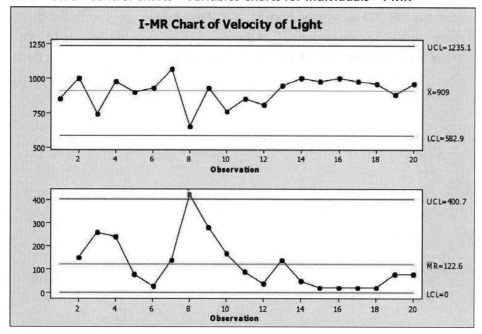

Test Results for MR Chart of Ex6.59Vel

```
TEST 1. One point more than 3.00 standard deviations from center line.
Test Failed at points:  8
```

The out-of-control signal on the moving range chart indicates a significantly large difference between successive measurements (7 and 8). Since neither of these measurements seems unusual, use all data for control limits calculations.

There may also be an early indication of less variability in the later measurements. For now, consider the process to be in a state of statistical process control.

6.71.

The uniformity of a silicon wafer following an etching process is determined by measuring the layer thickness at several locations and expressing uniformity as the range of the thicknesses. Table 6E.29 presents uniformity determinations for 30 consecutive wafers processed through the etching tool.

■ TABLE 6E.29
Uniformity Data for Exercise 6.71

Wafer	Uniformity	Wafer	Uniformity
1	11	16	15
2	16	17	16
3	22	18	12
4	14	19	11
5	34	20	18
6	22	21	14
7	13	22	13
8	11	23	18
9	6	24	12
10	11	25	13
11	11	26	12
12	23	27	15
13	14	28	21
14	12	29	21
15	7	30	14

(a) Is there evidence that uniformity is normally distributed? If not, find a suitable transformation for the data.

MTB > Stat > Basic Statistics > Normality Test

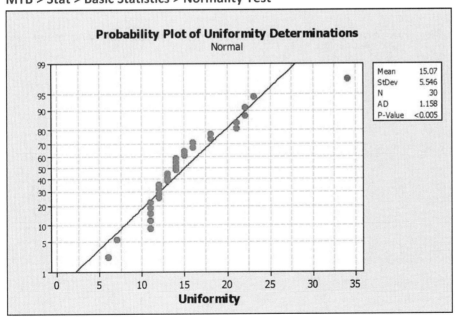

6.71.(a) continued

The data are not normally distributed, as evidenced by the "S"- shaped curve to the plot points on a normal probability plot, as well as the Anderson-Darling test p-value.

The data are skewed right, so a compressive transform such as natural log or square-root may be appropriate.

MTB > Stat > Basic Statistics > Normality Test

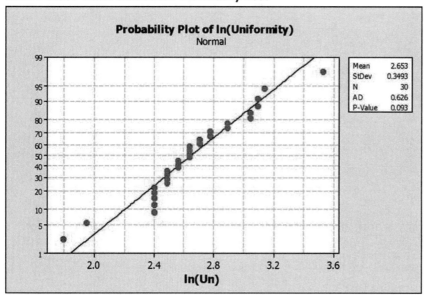

The distribution of the natural-log transformed uniformity measurements is approximately normally distributed.

6.71. continued

(b) Construct a control chart for individuals and a moving range control chart for uniformity for the etching process. Is the process in statistical control?

MTB > Stat > Control Charts > Variables Charts for Individuals > I-MR

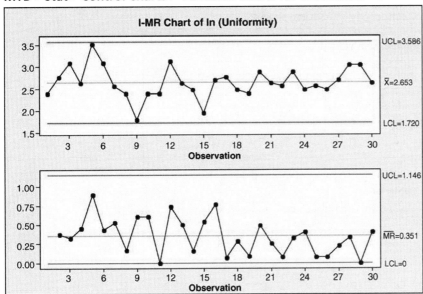

The etching process appears to be in statistical control.

6.73.

Reconsider the situation in Exercise 6.61. Construct an individuals control chart using the median of the span-two moving ranges to estimate variability. Compare this control chart to the one constructed in Exercise 6.61 and discuss.

MTB > Stat > Control Charts > Variables Charts for Individuals > I-MR
Select "Estimate" to change the method of estimating sigma

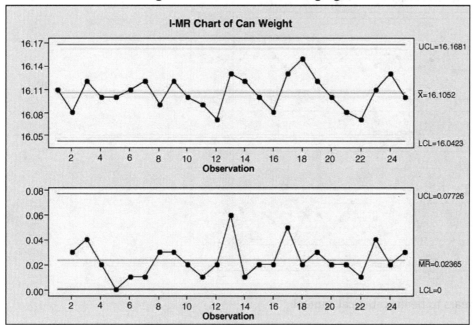

There is no difference between this chart and the one in Exercise 6.53; control limits for both are essentially the same.

6.75.

Reconsider the polymer viscosity data in Exercise 6.63. Use the median of the span-two moving ranges to estimate σ and set up the individuals control chart. Compare this chart to the one originally constructed using the average moving range method to estimate σ.

MTB > Stat > Control Charts > Variables Charts for Individuals > I-MR
Select "Estimate" to change the method of estimating sigma

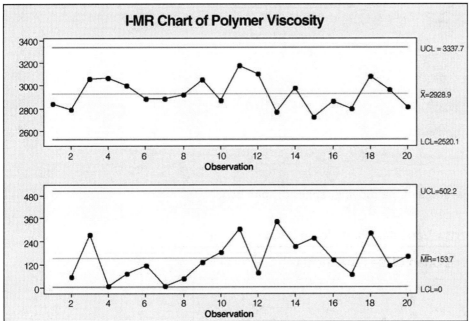

The median moving range method gives slightly wider control limits for both the Individual and Moving Range charts, with no practical meaning for this set of observations.

6.77.

Consider the individuals measurement data shown inTable6E.31.

■ **TABLE 6E.31**
Data for Exercise 6.77

Observation	x	Observation	x
1	10.07	14	9.58
2	10.47	15	8.80
3	9.45	16	12.94
4	9.44	17	10.78
5	8.99	18	11.26
6	7.74	19	9.48
7	10.63	20	11.28
8	9.78	21	12.54
9	9.37	22	11.48
10	9.95	23	13.26
11	12.04	24	11.10
12	10.93	25	10.82
13	11.54		

(a) Estimate σ using the average of the moving ranges of span two.

MTB > Stat > Control Charts > Variables Charts for Individuals > I-MR

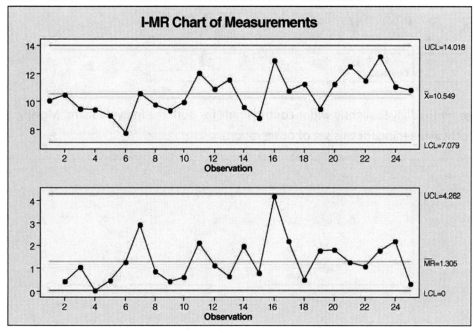

$$\hat{\sigma}_x = \overline{R} / d_2 = 1.305 / 1.128 = 1.157$$

6.77. continued

(b) Estimate σ using s/c_4.

MTB > Stat > Basic Statistics > Descriptive Statistics

Descriptive Statistics: Ex6.67Meas

```
              Total
Variable      Count     Mean   StDev   Median
Ex6.67Meas       25   10.549   1.342   10.630
```

$$\hat{\sigma}_x = S/c_4 = 1.342/0.7979 = 1.682$$

(c) Estimate σ using the median of the span-two moving ranges.

MTB > Stat > Control Charts > Variables Charts for Individuals > I-MR

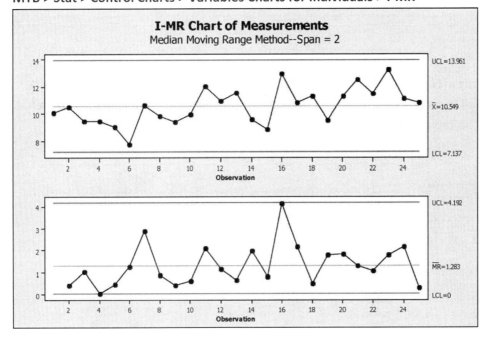

$$\hat{\sigma}_x = \overline{R}/d_2 = 1.283/1.128 = 1.137$$

6.77. continued

(d) Estimate σ using the average of the moving ranges of span 3, 4, ..., 20.

Average MR3 Chart: $\hat{\sigma}_x = \overline{R}/d_2 = 2.049/1.693 = 1.210$

Average MR4 Chart: $\hat{\sigma}_x = \overline{R}/d_2 = 2.598/2.059 = 1.262$

...

Average MR19 Chart: $\hat{\sigma}_x = \overline{R}/d_2 = 5.186/3.689 = 1.406$

Average MR20 Chart: $\hat{\sigma}_x = \overline{R}/d_2 = 5.36/3.735 = 1.435$

(e) Discuss the results you have obtained.

As the span of the moving range is increased, there are fewer observations to estimate the standard deviation, and the estimate becomes less reliable. For this example, σ gets larger as the span increases. This tends to be true for unstable processes.

6.79.

The diameter of the casting in Figure 6.25 is also an important quality characteristic. A coordinate measuring machine is used to measure the diameter of each casting at five different locations. Data for 20 casting are shown in the Table 6E.33.

■ **TABLE 6E.33**
Diameter Data for Exercise 6.79

Casting	Diameter 1	2	3	4	5
1	11.7629	11.7403	11.7511	11.7474	11.7374
2	11.8122	11.7506	11.7787	11.7736	11.8412
3	11.7742	11.7114	11.7530	11.7532	11.7773
4	11.7833	11.7311	11.7777	11.8108	11.7804
5	11.7134	11.6870	11.7305	11.7419	11.6642
6	11.7925	11.7611	11.7588	11.7012	11.7611
7	11.6916	11.7205	11.6958	11.7440	11.7062
8	11.7109	11.7832	11.7496	11.7496	11.7318
9	11.7984	11.8887	11.7729	11.8485	11.8416
10	11.7914	11.7613	11.7356	11.7628	11.7070
11	11.7260	11.7329	11.7424	11.7645	11.7571
12	11.7202	11.7537	11.7328	11.7582	11.7265
13	11.8356	11.7971	11.8023	11.7802	11.7903
14	11.7069	11.7112	11.7492	11.7329	11.7289
15	11.7116	11.7978	11.7982	11.7429	11.7154
16	11.7165	11.7284	11.7571	11.7597	11.7317
17	11.8022	11.8127	11.7864	11.7917	11.8167
18	11.7775	11.7372	11.7241	11.7773	11.7543
19	11.7753	11.7870	11.7574	11.7620	11.7673
20	11.7572	11.7626	11.7523	11.7395	11.7884

6.79. continued

(a) Set up $\bar{x}$ and R charts for this process, assuming the measurements on each casting form a rational subgroup.

MTB > Stat > Control Charts > Variables Charts for Subgroups > Xbar-R

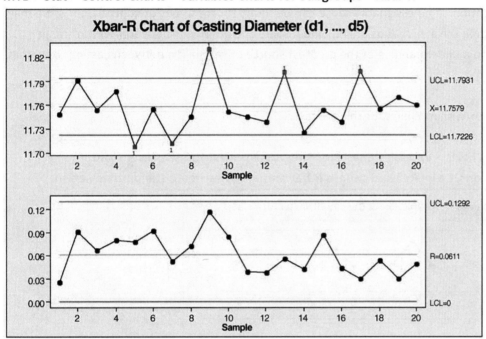

Xbar-R Chart of Ex6.69d1, ..., Ex6.69d5

Test Results for Xbar Chart of Ex6.69d1, ..., Ex6.69d5

```
TEST 1. One point more than 3.00 standard deviations from center line.
Test Failed at points:  5, 7, 9, 13, 17
TEST 5. 2 out of 3 points more than 2 standard deviations from center line (on
        one side of CL).
Test Failed at points:  7
```

6.79. continued

(b) Discuss the charts you have constructed in part (a).

Though the R chart is in control, plot points on the $\bar{x}$ chart bounce below and above the control limits. Since these are high precision castings, we might expect that the diameter of a single casting will not change much with location. If no assignable cause can be found for these out-of-control points, we may want to consider treating the averages as an Individual value and graphing "between/within" range charts. This will lead to a understanding of the greatest source of variability, between castings or within a casting.

(c) Construct "between/within" charts for this process.

MTB > Stat > Control Charts > Variables Charts for Subgroups > I-MR-R/S (Between/Within)
Select "I-MR-R/S Options, Estimate" and choose R-bar method to estimate standard deviation

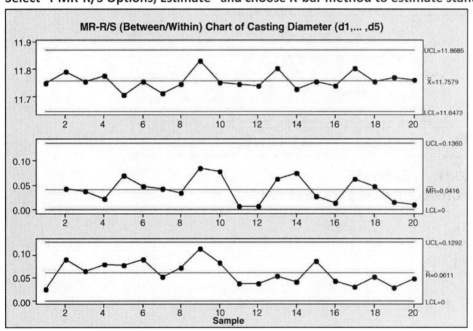

I-MR-R/S (Between/Within) Chart of Ex6.69d1, ..., Ex6.69d5
I-MR-R/S Standard Deviations of Ex6.69d1, ..., Ex6.69d5
```
Standard Deviations
Between          0.0349679
Within           0.0262640
Between/Within   0.0437327
```

6.79. continued

(d) Do you believe that the charts in part (c) are more informative than those in part (a)? Discuss why.

Yes, the charts in (c) are more informative than those in (a). The numerous out-of-control points on the chart in (a) result from using the wrong source of variability to estimate sample variance. Recall that this is a cast part, and that multiple diameter measurements of any single part are likely to be very similar (text p. 246). With the three charts in (c), the correct variance estimate is used for each chart. In addition, more clear direction is provided by the within and between charts for trouble-shooting out-of-control signals.

 (e) Provide a practical interpretation of the "within" chart.

The "within" chart is the usual R chart ($n > 1$). It describes the measurement variability within a sample (variability in diameter of a single casting). Though the nature of this process leads us to believe that the diameter at any location on a single casting does not change much, we should continue to monitor "within" to look for wear, damage, etc., in the wax mold.

6.81.

Consider the situation described in Exercise 6.80. A critical dimension (measured in μm) is of interest to the process engineer. Suppose that five fixed positions are used on each wafer (position 1 is the center) and that two consecutive wafers are selected from each batch. The data that result from several batches are shown in Table 6E.34.

■ **TABLE 6E.34**
Data for Exercise 6.81

Lot Number	Wafer Number	Position 1	2	3	4	5	Lot Number	Wafer Number	Position 1	2	3	4	5
1	1	2.15	2.13	2.08	2.12	2.10	11	1	2.15	2.13	2.14	2.09	2.08
	2	2.13	2.10	2.04	2.08	2.05		2	2.11	2.13	2.10	2.14	2.10
2	1	2.02	2.01	2.06	2.05	2.08	12	1	2.03	2.06	2.05	2.01	2.00
	2	2.03	2.09	2.07	2.06	2.04		2	2.04	2.08	2.03	2.10	2.07
3	1	2.13	2.12	2.10	2.11	2.08	13	1	2.05	2.03	2.05	2.09	2.08
	2	2.03	2.08	2.03	2.09	2.07		2	2.08	2.01	2.03	2.04	2.10
4	1	2.04	2.01	2.10	2.11	2.09	14	1	2.08	2.04	2.05	2.01	2.08
	2	2.07	2.14	2.12	2.08	2.09		2	2.09	2.11	2.06	2.04	2.05
5	1	2.16	2.17	2.13	2.18	2.10	15	1	2.14	2.13	2.10	2.10	2.08
	2	2.17	2.13	2.10	2.09	2.13		2	2.13	2.10	2.09	2.13	2.15
6	1	2.04	2.06	1.97	2.10	2.08	16	1	2.06	2.08	2.05	2.03	2.09
	2	2.03	2.10	2.05	2.07	2.04		2	2.03	2.01	1.99	2.06	2.05
7	1	2.04	2.02	2.01	2.00	2.05	17	1	2.05	2.03	2.08	2.01	2.04
	2	2.06	2.04	2.03	2.08	2.10		2	2.06	2.05	2.03	2.05	2.00
8	1	2.13	2.10	2.10	2.15	2.13	18	1	2.03	2.08	2.04	2.00	2.03
	2	2.10	2.09	2.13	2.14	2.11		2	2.04	2.03	2.05	2.01	2.04
9	1	1.95	2.03	2.08	2.07	2.08	19	1	2.16	2.13	2.10	2.13	2.12
	2	2.01	2.03	2.06	2.05	2.04		2	2.13	2.15	2.18	2.19	2.13
10	1	2.04	2.08	2.09	2.10	2.01	20	1	2.06	2.03	2.04	2.09	2.10
	2	2.06	2.04	2.07	2.04	2.01		2	2.01	1.98	2.05	2.08	2.06

6.81. continued

(a) What can you say about overall process capability?

MTB > Stat > Basic Statistics > Normality Test

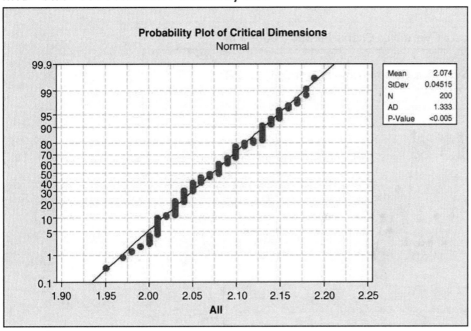

Although the p-value is very small, the plot points do fall along a straight line, with many repeated values. The wafer critical dimension is approximately normally distributed. The natural tolerance limits (± 3 sigma above and below mean) are:

$\bar{x} = 2.074, s = 0.04515$

$\text{UNTL} = \bar{x} + 3s = 2.074 + 3(0.04515) = 2.209; \quad \text{LNTL} = \bar{x} - 3s = 2.074 - 3(0.04515) = 1.939$

6.81. continued

(b) Can you construct control charts that allow within-wafer variability to be evaluated?
To evaluate within-wafer variability, construct an R chart for each sample of 5 wafer positions (two wafers per lot number), for a total of 40 subgroups.

MTB > Stat > Control Charts > Variables Charts for Subgroups > R

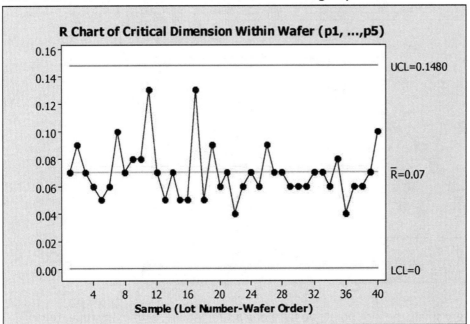

The Range chart is in control, indicating that within-wafer variability is also in control.

6.81. continued

(c) What control charts would you establish to evaluate variability between wafers? Set up these charts and use them to draw conclusions about the process.

To evaluate variability between wafers, set up Individuals and Moving Range charts where the x statistic is the average wafer measurement and the moving range is calculated between two wafer averages.

MTB > Stat > Control Charts > Variables Charts for Subgroups > I-MR-R/S (Between/Within)
Select "I-MR-R/S Options, Estimate" and choose R-bar method to estimate standard deviation

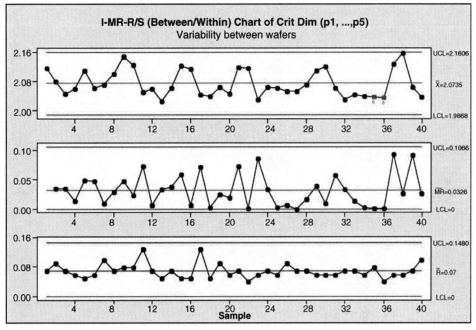

I-MR-R/S Standard Deviations of Ex6.71p1, ..., Ex6.71p5

```
Standard Deviations
Between          0.0255911
Within           0.0300946
Between/Within   0.0395043
```

Both "between" control charts (Individuals and Moving Range) are in control, indicating that between-wafer variability is also in-control. The "within" chart (Range) is not required to evaluate variability between wafers.

6.81. continued

(d) What control charts would you use to evaluate lot-to-lot variability? Set up these charts and use them to draw conclusions about lot-to-lot variability.

To evaluate lot-to-lot variability, three charts are needed: (1) lot average, (2) moving range between lot averages, and (3) range within a lot—the Minitab "between/within" control charts.

MTB > Stat > Control Charts > Variables Charts for Subgroups > I-MR-R/S (Between/Within)

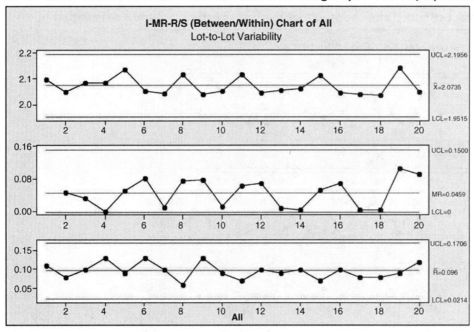

I-MR-R/S Standard Deviations of Ex6.71All
```
Standard Deviations
Between        0.0394733
Within         0.0311891
Between/Within 0.0503081
```

All three control charts are in control, indicating that the lot-to-lot variability is also in-control.

Chapter 7

Control Charts for Attributes

LEARNING OBJECTIVES

After completing this chapter you should be able to:
1. Understand the statistical basis of attributes control charts
2. Know how to design attributes control charts
3. Know how to set up and use the p chart for fraction nonconforming
4. Know how to set up and use the np control chart for the number of nonconforming items
5. Know how to set up and use the c control chart for defects
6. Know how to set up and use the u control chart for defects per unit
7. Use attributes control charts with variable sample size
8. Understand the advantages and disadvantages of attributes versus variables control charts
9. Understand the rational subgroup concept for attributes control charts
10. Determine the average run length for attributes control charts

IMPORTANT TERMS AND CONCEPTS

Attribute data

Average run length for attributes control charts

Cause-and-effect diagram

Choice between attributes and variables data

Control chart for defects or nonconformities per unit or u chart

Control chart for fraction nonconforming or p chart

Control chart for nonconformities or c chart

Control chart for number nonconforming or np chart

Defect

Defective

Demerit systems for attribute data

Design of attributes control charts

Fraction defective

Fraction nonconforming

Nonconformity

Operating characteristic curve for the c and u charts

Operating characteristic curve for the p chart

Pareto chart

Standardized control charts

Time between occurrence control charts

Variable sample size for attributes control chart

EXERCISES

Minitab® Notes:
1. The Minitab convention for determining whether a point is out of control is: (1) if a plot point is within the control limits, it is in control, or (2) if a plot point is on or beyond the limits, it is out of control.
2. Minitab defines some sensitizing rules for control charts differently than the standard rules. In particular, a run of n consecutive points on one side of the center line is defined as 9 points, not 8. This can be changed in dialog boxes, or under Tools > Options > Control Charts and Quality Tools > Tests.

7.1.

A financial services company monitors loan applications. Every day 50 applications are assessed for the accuracy of the information on the form. Results for 20 days are $\sum_{i=1}^{20} D_i = 46$ where D_i is the number of loans on the ith day that re determined to have at least one error. What are the center line and control limits on the fraction nonconforming control chart?

$$n = 50; \quad m = 20; \quad \sum_{i=1}^{20} D_i = 46; \quad \bar{p} = \frac{\sum_{i=1}^{m} D_i}{mn} = \frac{46}{20(50)} = 0.046$$

$$\text{UCL}_p = \bar{p} + 3\sqrt{\frac{\bar{p}(1-\bar{p})}{n}} = 0.046 + 3\sqrt{\frac{0.046(1-0.046)}{50}} = 0.046 + 0.089 = 0.135$$

$$\text{LCL}_p = \bar{p} - 3\sqrt{\frac{\bar{p}(1-\bar{p})}{n}} = 0.046 - 3\sqrt{\frac{0.046(1-0.046)}{50}} = 0.046 - 0.089 \Rightarrow 0$$

7.9.

The data in Table 7E.3 give the number of nonconforming bearing and seal assemblies in samples of size 100. Construct a fraction nonconforming control chart for these data. If any points plot out of control, assume that assignable causes can be found and determine the revised control limits.

■ **TABLE 7E.3**
Data for Exercise 7.9

Sample Number	Number of Nonconforming Assemblies	Sample Number	Number of Nonconforming Assemblies
1	7	11	6
2	4	12	15
3	1	13	0
4	3	14	9
5	6	15	5
6	8	16	1
7	10	17	4
8	5	18	5
9	2	19	7
10	7	20	12

$$n = 100; \quad m = 20; \quad \sum_{i=1}^{m} D_i = 117; \quad \bar{p} = \frac{\sum_{i=1}^{m} D_i}{mn} = \frac{117}{20(100)} = 0.0585$$

$$\text{UCL}_p = \bar{p} + 3\sqrt{\frac{\bar{p}(1-\bar{p})}{n}} = 0.0585 + 3\sqrt{\frac{0.0585(1-0.0585)}{100}} = 0.1289$$

$$\text{LCL}_p = \bar{p} - 3\sqrt{\frac{\bar{p}(1-\bar{p})}{n}} = 0.0585 - 3\sqrt{\frac{0.0585(1-0.0585)}{100}} = 0.0585 - 0.0704 \Rightarrow 0$$

7.9. continued

MTB > Stat > Control Charts > Attributes Charts > P
For Subgroup sizes, enter 100

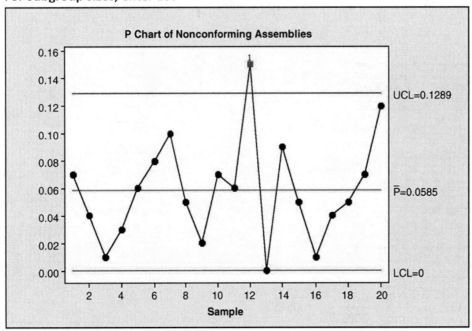

Test Results for P Chart of Ex7.1Num
```
TEST 1. One point more than 3.00 standard deviations from center line.
Test Failed at points:   12
```

Sample 12 is out-of-control, so remove from control limit calculation:

$$n=100; \quad m=19; \quad \sum_{i=1}^{m} D_i = 102; \quad \bar{p} = \frac{\sum_{i=1}^{m} D_i}{mn} = \frac{102}{19(100)} = 0.0537$$

$$UCL_p = 0.0537 + 3\sqrt{\frac{0.0537(1-0.0537)}{100}} = 0.1213$$

$$LCL_p = 0.0537 - 3\sqrt{\frac{0.0537(1-0.0537)}{100}} = 0.0537 - 0.0676 \Rightarrow 0$$

7.9. continued

MTB > Stat > Control Charts > Attributes Charts > P

For Subgroup sizes, enter 100

To exclude Sample 12, select P Chart Options, Estimate tab, and Omit subgroup 16

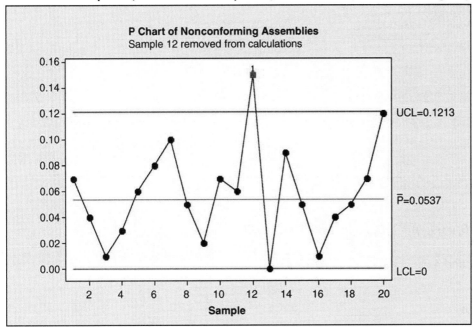

Test Results for P Chart of Ex7.1Num

```
TEST 1. One point more than 3.00 standard deviations from center line.
Test Failed at points:  12
```

No additional samples signal out of control. Use UCL = 0.1213, CL = 0.0537 and LCL = 0 to monitor the process.

7.11.

The data in Table 7E.5 represent the results of inspecting all units of a personal computer produced for the last 10 days. Does the process appear to be in control?

■ TABLE 7E.5
Personal Computer Inspecting Results for Exercise 7.11

Day	Units Inspected	Nonconforming Units	Fraction Nonconforming
1	80	4	0.050
2	110	7	0.064
3	90	5	0.056
4	75	8	0.107
5	130	6	0.046
6	120	6	0.050
7	70	4	0.057
8	125	5	0.040
9	105	8	0.076
10	95	7	0.074

$$m=10; \quad \sum_{i=1}^{m} n_i = 1000; \quad \sum_{i=1}^{m} D_i = 60; \quad \bar{p} = \sum_{i=1}^{m} D_i \Big/ \sum_{i=1}^{m} n_i = 60/1000 = 0.06$$

$$\text{UCL}_i = \bar{p} + 3\sqrt{\bar{p}(1-\bar{p})/n_i} \quad \text{and} \quad \text{LCL}_i = \max\{0, \bar{p} - 3\sqrt{\bar{p}(1-\bar{p})/n_i}\}$$

As an example, for $n = 80$:

$$\text{UCL}_1 = \bar{p} + 3\sqrt{\bar{p}(1-\bar{p})/n_1} = 0.06 + 3\sqrt{0.06(1-0.06)/80} = 0.1397$$

$$\text{LCL}_1 = \bar{p} - 3\sqrt{\bar{p}(1-\bar{p})/n_1} = 0.06 - 3\sqrt{0.06(1-0.06)/80} = 0.06 - 0.0797 \Rightarrow 0$$

7.11. continued

MTB > Stat > Control Charts > Attributes Charts > P

For Subgroup sizes, enter column with Units Inspected

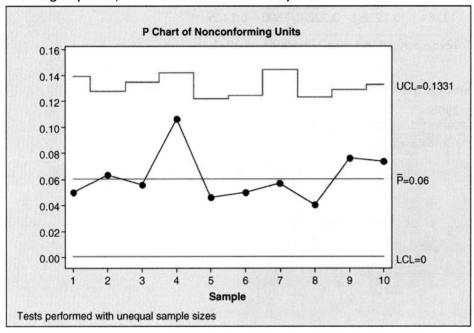

The process appears to be in statistical control.

7.13.

A process produces rubber belts in lots of size 2500. Inspection records on the last 20 lots reveal the data in Table 7E.7.

■ TABLE 7E.7
Inspection Data for Exercise 7.13

Lot Number	Number of Nonconforming Belts	Lot Number	Number of Nonconforming Belts
1	230	11	456
2	435	12	394
3	221	13	285
4	346	14	331
5	230	15	198
6	327	16	414
7	285	17	131
8	311	18	269
9	342	19	221
10	308	20	407

7.13. continued

(a) Compute trial control limits for a fraction nonconforming control chart.

$$UCL = \bar{p} + 3\sqrt{\bar{p}(1-\bar{p})/n} = 0.1228 + 3\sqrt{0.1228(1-0.1228)/2500} = 0.1425$$
$$LCL = \bar{p} - 3\sqrt{\bar{p}(1-\bar{p})/n} = 0.1228 - 3\sqrt{0.1228(1-0.1228)/2500} = 0.1031$$

MTB > Stat > Control Charts > Attributes Charts > P
For Subgroup sizes, enter 2500

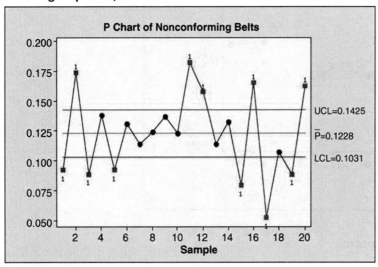

Test Results for P Chart of Ex7.5Num
```
TEST 1. One point more than 3.00 standard deviations from center line.
Test Failed at points:   1, 2, 3, 5, 11, 12, 15, 16, 17, 19, 20
```

(b) If you wanted to set up a control chart for controlling future production, how would you use these data to obtain the center line and control limits for the chart?

So many subgroups are out of control (11 of 20) that the data should not be used to establish control limits for future production. Instead, the process should be investigated for causes of the wild swings in p.

7.15.

A control chart indicates that the current process fraction nonconforming is 0.02. If 50 items are inspected each day, what is the probability of detecting a shift in the fraction nonconforming to 0.04 on the first day after the shift? By the end of the third day following the shift?

$\bar{p} = 0.02; n = 50$

$\text{UCL} = \bar{p} + 3\sqrt{\bar{p}(1-\bar{p})/n} = 0.02 + 3\sqrt{0.02(1-0.02)/50} = 0.0794$

$\text{LCL} = \bar{p} - 3\sqrt{\bar{p}(1-\bar{p})/n} = 0.02 - 3\sqrt{0.02(1-0.02)/50} = 0.02 - 0.0594 \Rightarrow 0$

Since $p_{new} = 0.04 < 0.1$ and $n = 50$ is "large", use the Poisson approximation to the binomial with $\lambda = np_{new} = 50(0.04) = 2.00$.

Pr{detect|shift}

$\qquad = 1 - \text{Pr\{not detect|shift\}}$

$\qquad = 1 - \beta$

$\qquad = 1 - [\text{Pr}\{D < n\text{UCL} \mid \lambda\} - \text{Pr}\{D \le n\text{LCL} \mid \lambda\}]$

$\qquad = 1 - \text{Pr}\{D < 50(0.0794) \mid 2\} + \text{Pr}\{D \le 50(0) \mid 2\}$

$\qquad = 1 - \text{POI}(3,2) + \text{POI}(0,2) = 1 - 0.857 + 0.135 = 0.278$

where POI(·) is the cumulative Poisson distribution.

Pr{detected by 3rd sample} = 1 − Pr{detected after 3rd} = $1 - (1 - 0.278)^3 = 0.624$

7.17.

Diodes used on printed circuit boards are produced in lots of size 1000. We wish to control the process producing these diodes by taking samples of size 64 from each lot. If the nominal value of the fraction nonconforming is $p = 0.10$, determine the parameters of the appropriate control chart. To what level must the fraction nonconforming increase to make the β-risk equal to 0.50? What is the minimum sample size that would give a positive lower control limit for this chart?

$\bar{p} = 0.10; n = 64$

$\text{UCL} = \bar{p} + 3\sqrt{\bar{p}(1-\bar{p})/n} = 0.10 + 3\sqrt{0.10(1-0.10)/64} = 0.2125$

$\text{LCL} = \bar{p} - 3\sqrt{\bar{p}(1-\bar{p})/n} = 0.10 - 3\sqrt{0.10(1-0.10)/64} = 0.10 - 0.1125 \Rightarrow 0$

$\beta = \Pr\{D < n\text{UCL} \mid p\} - \Pr\{D \le n\text{LCL} \mid p\}$

$\quad = \Pr\{D < 64(0.2125) \mid p\} - \Pr\{D \le 64(0) \mid p\}$

$\quad = \Pr\{D < 13.6) \mid p\} - \Pr\{D \le 0 \mid p\}$

p	Pr{D ≤ 13|p}	Pr{D ≤ 0|p}	β
0.05	0.999999	0.037524	0.962475
0.10	0.996172	0.001179	0.994993
0.20	0.598077	0.000000	0.598077
0.21	**0.519279**	**0.000000**	**0.519279**
0.22	0.44154	0.000000	0.44154
0.215	0.480098	0.000000	0.480098
0.212	0.503553	0.000000	0.503553

Assuming $L = 3$ sigma control limits,

$n > \dfrac{(1-p)}{p} L^2$

$\quad > \dfrac{(1-0.10)}{0.10}(3)^2$

$\quad > 81$

7.19.

A control chart for the fraction nonconforming is to be established using a center line of p = 0.10. What sample size is required if we wish to detect a shift in the process fraction nonconforming to 0.20 with probability 0.50?

$p = 0.10$; $p_{new} = 0.20$; desire Pr{detect} = 0.50; assume $k = 3$ sigma control limits

$\delta = p_{new} - p = 0.20 - 0.10 = 0.10$

$$n = \left(\frac{k}{\delta}\right)^2 p(1-p) = \left(\frac{3}{0.10}\right)^2 (0.10)(1-0.10) = 81$$

7.21.

A process is being controlled with a fraction nonconforming control chart. The process average has been shown to be 0.07. Three-sigma control limits are used, and the procedure calls for taking daily samples of 400 items.

(a) Calculate the upper and lower control limits.

$\bar{p} = 0.07$; $k = 3$ sigma control limits; $n = 400$

$UCL = \bar{p} + 3\sqrt{p(1-p)/n} = 0.07 + 3\sqrt{0.07(1-0.07)/400} = 0.108$

$LCL = \bar{p} - 3\sqrt{p(1-p)/n} = 0.07 - 3\sqrt{0.07(1-0.07)/400} = 0.032$

(b) If the process average should suddenly shift to 0.10, what is the probability that the shift would be detected on the first subsequent sample?

np_{new} = 400(0.10) = > 40, so use the normal approximation to the binomial.

Pr{detect on 1st sample} $= 1 - $Pr{not detect on 1st sample}

$$= 1 - \beta$$
$$= 1 - [\text{Pr}\{\hat{p} < UCL \mid p\} - \text{Pr}\{\hat{p} \le LCL \mid p\}]$$
$$= 1 - \Phi\left(\frac{UCL - p}{\sqrt{p(1-p)/n}}\right) + \Phi\left(\frac{LCL - p}{\sqrt{p(1-p)/n}}\right)$$
$$= 1 - \Phi\left(\frac{0.108 - 0.1}{\sqrt{0.1(1-0.1)/400}}\right) + \Phi\left(\frac{0.032 - 0.1}{\sqrt{0.1(1-0.1)/400}}\right)$$
$$= 1 - \Phi(0.533) + \Phi(-4.533)$$
$$= 1 - 0.703 + 0.000$$
$$= 0.297$$

7.21. continued

(c) What is the probability that the shift in part (b) would be detect on the first or second sample taken after the shift?

Pr{detect on 1st or 2nd sample}

$\qquad$ = Pr{detect on 1st} + Pr{not on 1st}×Pr{detect on 2nd}

$\qquad$ = 0.297 + (1 − 0.297)(0.297)

$\qquad$ = 0.506

7.23.

A control chart is used to control the fraction nonconforming for a plastic part manufactured in an injection molding process. Ten subgroups yield the data in Table 7E.9.

■ TABLE 7E.9
Nonconforming Unit Data for Exercise 7.23

Sample Number	Sample Size	Number Nonconforming
1	100	10
2	100	15
3	100	31
4	100	18
5	100	24
6	100	12
7	100	23
8	100	15
9	100	8
10	100	8

(a) Set up a control chart for the number nonconforming in samples of $n = 100$.

$$m = 10; \quad n = 100; \quad \sum_{i=1}^{10} D_i = 164; \quad \bar{p} = \sum_{i=1}^{10} D_i \Big/ (mn) = 164 / \left[10(100)\right] = 0.164; \quad n\bar{p} = 16.4$$

$$\text{UCL} = n\bar{p} + 3\sqrt{n\bar{p}(1-\bar{p})} = 16.4 + 3\sqrt{16.4(1-0.164)} = 27.51$$

$$\text{LCL} = n\bar{p} - 3\sqrt{n\bar{p}(1-\bar{p})} = 16.4 - 3\sqrt{16.4(1-0.164)} = 5.292$$

7.23.(a) continued

MTB > Stat > Control Charts > Attributes Charts > NP

For Subgroup sizes, enter 100

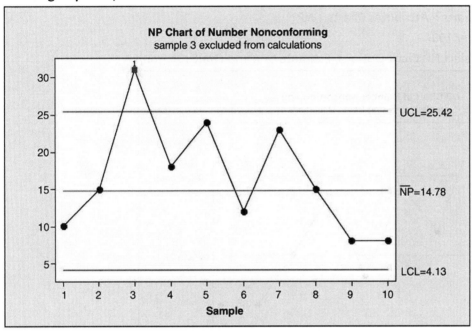

Test Results for NP Chart of Ex7.15Num

```
TEST 1. One point more than 3.00 standard deviations from center line.
Test Failed at points:  3
```

7.23.(a) continued

Recalculate control limits without sample 3:

MTB > Stat > Control Charts > Attributes Charts > NP
For Subgroup sizes, enter 100
To exclude Sample 3, select NP Chart Options, Estimate tab, and Omit subgroup 3

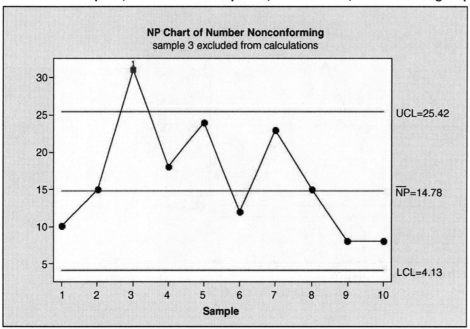

Test Results for NP Chart of Ex7.15Num
```
TEST 1. One point more than 3.00 standard deviations from center line.
Test Failed at points:  3
```

Recommend using control limits from second chart (calculated less sample 3), with center line = 14.78, UCL = 25.42 and LCL = 4.13.

7.23. continued

(b) For the chart established in part (a), what is the probability of detecting a shift in the process fraction nonconforming to 0.30 on the first sample after the shift has occurred?

p_{new} = 0.30. Since p = 0.30 is not too far from 0.50, and n = 100 > 10, the normal approximation to the binomial can be used.

$$
\begin{aligned}
\Pr\{\text{detect on 1st}\} &= 1 - \Pr\{\text{not detect on 1st}\} \\
&= 1 - \beta \\
&= 1 - [\Pr\{D < UCL \mid p\} - \Pr\{D \le LCL \mid p\}] \\
&= 1 - \Phi\left(\frac{UCL + 1/2 - np}{\sqrt{np(1-p)}}\right) + \Phi\left(\frac{LCL - 1/2 - np}{\sqrt{np(1-p)}}\right) \\
&= 1 - \Phi\left(\frac{25.42 + 0.5 - 30}{\sqrt{30(1-0.3)}}\right) + \Phi\left(\frac{4.13 - 0.5 - 30}{\sqrt{30(1-0.3)}}\right) \\
&= 1 - \Phi(-0.8903) + \Phi(-5.7544) \\
&= 1 - (0.187) + (0.000) \\
&= 0.813
\end{aligned}
$$

7.25.

(a) A control chart for the number nonconforming is to be established, based on samples of size 400. To start the control chart, 30 samples were selected and the number nonconforming in each sample determined, yielding $\sum_{i=1}^{30} D_i = 1,200$. What are the parameters of the np chart?

$$\bar{p} = \sum_{i=1}^{m} D_i \Big/ (mn) = 1200/[30(400)] = 0.10; \quad n\bar{p} = 400(0.10) = 40$$

$$UCL_{np} = n\bar{p} + 3\sqrt{n\bar{p}(1-\bar{p})} = 40 + 3\sqrt{40(1-0.10)} = 58$$

$$LCL_{np} = n\bar{p} - 3\sqrt{n\bar{p}(1-\bar{p})} = 40 - 3\sqrt{40(1-0.10)} = 22$$

(b) Suppose the process average fraction nonconforming shifted to 0.15. What is the probability that the shift would be detected on the first subsequent sample?

np_{new} = 400 (0.15) = 60 > 15, so use the normal approximation to the binomial.

$$Pr\{detect\ on\ 1st\ sample\,|\,p\} = 1 - Pr\{not\ detect\ on\ 1st\ sample\,|\,p\}$$
$$= 1 - \beta$$
$$= 1 - [Pr\{D < UCL\,|\,np\} - Pr\{D \le LCL\,|\,np\}]$$
$$= 1 - \Phi\left(\frac{UCL + 1/2 - np}{\sqrt{np(1-p)}}\right) + \Phi\left(\frac{LCL - 1/2 - np}{\sqrt{np(1-p)}}\right)$$
$$= 1 - \Phi\left(\frac{58 + 0.5 - 60}{\sqrt{60(1-0.15)}}\right) + \Phi\left(\frac{22 - 0.5 - 60}{\sqrt{60(1-0.15)}}\right)$$
$$= 1 - \Phi(-0.210) + \Phi(-5.39)$$
$$= 1 - 0.417 + 0.000$$
$$= 0.583$$

7.27.

Consider the control chart designed in Exercise 7.25. Find the average run length to detect a shift to a fraction nonconforming of 0.15.

from 7.25.(b), $1 - \beta = 0.583$

$ARL_1 = 1/(1 - \beta) = 1/(0.583) = 1.715 \cong 2$

7.29.

A maintenance group improves the effectiveness of its repair work by monitoring the number of maintenance requests that require a second call to complete the repair. Twenty weeks of data are shown in Table 7E.10.

■ **TABLE 7E.10**
Data for Exercise 7.29

Week	Total Requests	Second Visit Required	Week	Total Requests	Second Visit Required
1	200	6	11	100	1
2	250	8	12	100	0
3	250	9	13	100	1
4	250	7	14	200	4
5	200	3	15	200	5
6	200	4	16	200	3
7	150	2	17	200	10
8	150	1	18	200	4
9	150	0	19	250	7
10	150	2	20	250	6

(a) Find trial control limits for this process.

For a p chart with variable sample size: $\bar{p} = \sum_i D_i / \sum_i n_i = 83/3750 = 0.0221$ and control limits are at

$$\bar{p} \pm 3\sqrt{\bar{p}(1-\bar{p})/n_i}$$

n_i	[LCL$_i$, UCL$_i$]
100	[0, 0.0662]
150	[0, 0.0581]
200	[0, 0.0533]
250	[0, 0.0500]

7.29.(a) continued

MTB > Stat > Control Charts > Attributes Charts > P
For Subgroup sizes, enter column with Total Requests

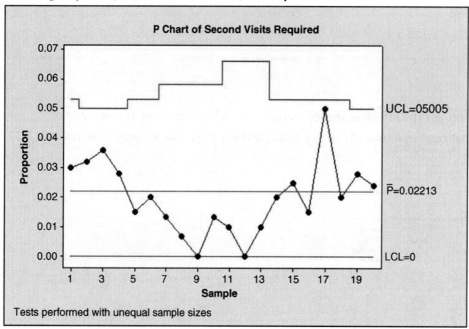

Process is in statistical control.

(b) Design a control chart for controlling future production.

There are two approaches for controlling future production. The first approach would be to plot $\hat{p}_i$ and use constant limits unless there is a different size sample or a plot point near a control limit. In those cases, calculate the exact control limits by $\bar{p} \pm 3\sqrt{\bar{p}(1-\bar{p})/n_i} = 0.0221 \pm 3\sqrt{0.0216/n_i}$. The second approach, preferred in many cases, would be to construct standardized control limits with control limits at ± 3, and to plot $Z_i = (\hat{p}_i - 0.0221)/\sqrt{0.0221(1-0.0221)/n_i}$.

7.31.

Construct a standardized control chart for the data in Exercise 7.29.

$$z_i = (\hat{p}_i - \bar{p})\Big/\sqrt{\bar{p}(1-\bar{p})/n_i} = (\hat{p}_i - 0.0221)\Big/\sqrt{0.0216/n_i}$$

MTB > Stat > Control Charts > Variables Charts for Individuals > Individuals
Under I Chart Options, on the Parameters tab, enter Mean = 0 and Standard deviation = 1

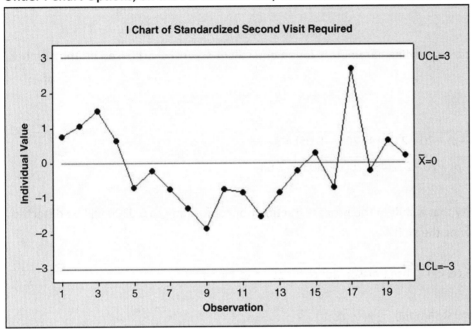

Process is in statistical control.

7.39.

A fraction nonconforming control chart with $n = 400$ has the following parameters: UCL = 0.0809, Center line = 0.0500, LCL = 0.0191.

(a) Find the width of the control limits in standard deviation units.

$$0.0809 = 0.05 + L\sqrt{0.05(1-0.05)/400} = 0.05 + L(0.0109)$$
$$L = 2.8349$$

(b) What would be the corresponding parameters for an equivalent control chart based on the number nonconforming?

$$CL = np = 400(0.05) = 20$$
$$UCL = np + 2.8349\sqrt{np(1-p)} = 20 + 2.8349\sqrt{20(1-0.05)} = 32.36$$
$$LCL = np - 2.8349\sqrt{np(1-p)} = 20 - 2.8349\sqrt{20(1-0.05)} = 7.64$$

(c) What is the probability that a shift in the process fraction nonconforming to 0.0300 will be detected on the first sample following the shift?

$n = 400$ is large and $p = 0.05 < 0.1$, use Poisson approximation to binomial.

Pr{detect shift to 0.03 on 1st sample}
$= 1 - \text{Pr\{not detect\}}$
$= 1 - \beta$
$= 1 - [\text{Pr}\{D < UCL \mid \lambda\} - \text{Pr}\{D \le LCL \mid \lambda\}]$
$= 1 - \text{Pr}\{D < 32.36 \mid 12\} + \text{Pr}\{D \le 7.64 \mid 12\}$
$= 1 - POI(32,12) + POI(7,12)$
$= 1 - 1.0000 + 0.0895$
$= 0.0895$

where $POI(\cdot)$ is the cumulative Poisson distribution.

7.41.

A fraction nonconforming control chart is to be established with a center line of 0.01 and two-sigma control limits.

(a) How large should the sample size be if the lower control limit is to be nonzero?

$$n > \left(\frac{1-p}{p}\right) L^2$$

$$> \left(\frac{1-0.01}{0.01}\right) 2^2$$

$$> 396$$

$$\geq 397$$

(b) How large should the sample size be if we wish the probability of detecting a shift to 0.04 to be 0.50?

$$\delta = 0.04 - 0.01 = 0.03$$

$$n = \left(\frac{L}{\delta}\right)^2 p(1-p) = \left(\frac{2}{0.03}\right)^2 (0.01)(1-0.01) = 44$$

7.43.

A process that produces bearing housings is controlled with a fraction nonconforming control chart, using sample size $n = 100$ and a center line $\bar{p} = 0.02$.

(a) Find the three-sigma limits for this chart.

$$UCL = \bar{p} + 3\sqrt{\bar{p}(1-\bar{p})/n} = 0.02 + 3\sqrt{0.02(1-0.02)/100} = 0.062$$

$$LCL = \bar{p} - 3\sqrt{\bar{p}(1-\bar{p})/n} = 0.02 - 3\sqrt{0.02(1-0.02)/100} \Rightarrow 0$$

7.43. continued

(b) Analyze the ten new samples ($n = 100$) shown in Table 7E.11 for statistical control. What conclusions can you draw about the process now?

■ **TABLE 7E.11**
Data for Exercise 7.43, part (b)

Sample Number	Number Nonconforming	Sample Number	Number Nonconforming
1	5	6	1
2	2	7	2
3	3	8	6
4	8	9	3
5	4	10	4

MTB > Stat > Control Charts > Attributes Charts > P

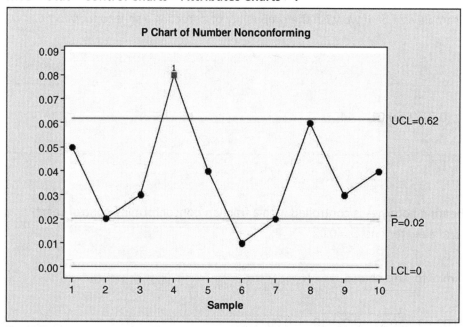

Test Results for P Chart of Ex7.31Num

```
TEST 1. One point more than 3.00 standard deviations from center line.
Test Failed at points:   4
```

Sample 4 exceeds the upper control limit.

$\bar{p} = 0.038$ and $\hat{\sigma}_p = 0.0191$

7.45.

Consider the fraction nonconforming control chart in Exercise 7.12. Find the equivalent *np* chart.

$n = 150$; $m = 20$; $\sum D = 50$; $\bar{p} = 0.0167$

$CL = n\bar{p} = 150(0.0167) = 2.505$

$UCL = n\bar{p} + 3\sqrt{n\bar{p}(1-\bar{p})} = 2.505 + 3\sqrt{2.505(1-0.0167)} = 7.213$

$LCL = n\bar{p} - 3\sqrt{n\bar{p}(1-\bar{p})} = 2.505 - 4.708 \Rightarrow 0$

MTB > Stat > Control Charts > Attributes Charts > NP

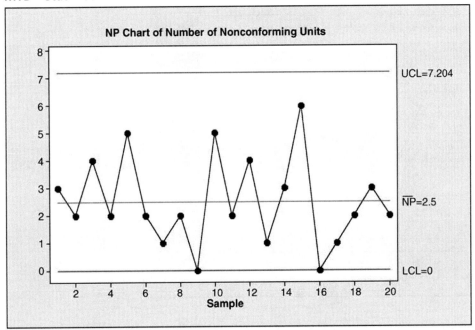

The process is in control; results are the same as for the *p* chart.

7.47.

Construct a standardized control chart for the data in Exercise 7.11.

$\bar{p} = 0.06$

$z_i = (\hat{p}_i - 0.06) \big/ \sqrt{0.06(1 - 0.06)/n_i} = (\hat{p}_i - 0.06) \big/ \sqrt{0.0564/n_i}$

MTB > Stat > Control Charts > Variables Charts for Individuals > Individuals
Select I Chart Options, Parameters tab, to enter a Mean of 0 and Standard deviation of 1

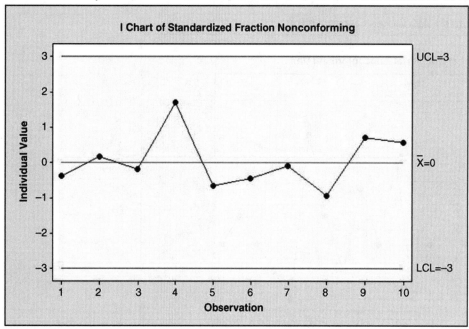

The process is in control; results are the same as for the *p* chart.

7.49.

A paper mill uses a control chart to monitor the imperfection nin finished rolls of paper. Production output is inspected for 20 days, and the resulting data are shown in Table 7E.13. Use these data to set up a control chart for nonconformities per roll of paper. Does the process appear to be in statistical control? What center line and control limits would you recommend for controlling current production?

■ TABLE 7E.13
Data on Imperfections in Rolls of Paper

Day	Number of Rolls Produced	Total Number of Imperfections	Day	Number of Rolls Produced	Total Number of Imperfections
1	18	12	11	18	18
2	18	14	12	18	14
3	24	20	13	18	9
4	22	18	14	20	10
5	22	15	15	20	14
6	22	12	16	20	13
7	20	11	17	24	16
8	20	15	18	24	18
9	20	12	19	22	20
10	20	10	20	21	17

$$CL = \bar{u} = 0.7007$$

$$UCL_i = \bar{u} + 3\sqrt{\bar{u}/n_i} = 0.7007 + 3\sqrt{0.7007/n_i}$$

$$LCL_i = \bar{u} - 3\sqrt{\bar{u}/n_i} = 0.7007 - 3\sqrt{0.7007/n_i}$$

n_i	$[LCL_i, UCL_i]$
18	[0.1088, 1.2926]
20	[0.1392, 1.2622]
21	[0.1527, 1.2487]
22	[0.1653, 1.2361]
24	[0.1881, 1.2133]

7.49. continued

MTB > Stat > Control Charts > Attributes Charts > U

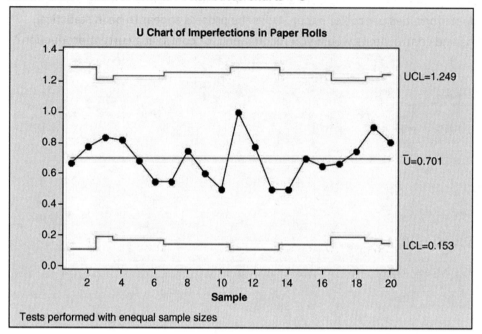

The process is in statistical control, with no patterns or out-of-control points. Use a control chart with CL = 0.701, UCL = 1.249, and LCL = 0.153 to control current production.

7.51.

Continuation of Exercise7.49. Consider the paper-making process in Exercise 7.49. Set up a standardized u chart for this process.

$$z_i = (u_i - \bar{u}) / \sqrt{\bar{u}/n_i} = (u_i - 0.7007) / \sqrt{0.7007 / n_i}$$

MTB > Stat > Control Charts > Variables Charts for Individuals > Individuals
Select I Chart Options, Parameters tab, to enter a Mean of 0 and Standard deviation of 1

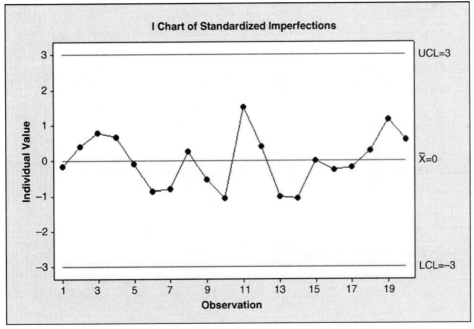

The process is in statistical control, with no patterns or out-of-control points.

7.53.

The data in Table 7E.15 represent the number of nonconformities per 1,000 meters in telephone cable. From analysis of these data, would you conclude that the process is in statistical control? What control procedure would you recommend for future production?

7.53. continued

■ **TABLE 7E.15**

Telephone Cable Data for Exercise 7.53

Sample Number	Number of Nonconformities	Sample Number	Number of Nonconformities
1	1	12	6
2	1	13	9
3	3	14	11
4	7	15	15
5	8	16	8
6	10	17	3
7	5	18	6
8	13	19	7
9	0	20	4
10	19	21	9
11	24	22	20

Utilize a *c* chart based on # of nonconformities per 1,000 meters.

$$CL = \bar{c} = 8.59; \quad UCL = \bar{c} + 3\sqrt{\bar{c}} = 8.59 + 3\sqrt{8.59} = 17.384; \quad LCL = \bar{c} - 3\sqrt{\bar{c}} = 8.59 - 3\sqrt{8.59} \Rightarrow 0$$

MTB > Stat > Control Charts > Attributes Charts > C

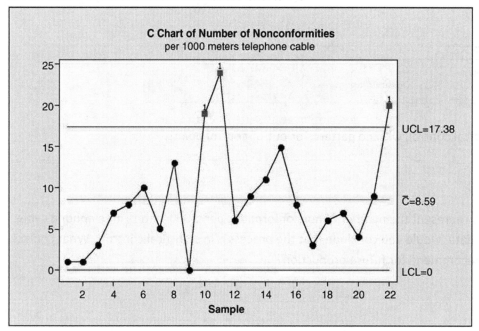

7.53. continued

Test Results for C Chart of Ex7.41Num
```
TEST 1. One point more than 3.00 standard deviations from center line.
Test Failed at points:  10, 11, 22
```

The process is not in statistical control; three subgroups exceed the UCL. Exclude subgroups 10, 11 and 22, and re-calculate the control limits. Subgroup 15 will then be out of control and should also be excluded.

$$CL = \bar{c} = 6.17$$
$$UCL = \bar{c} + 3\sqrt{\bar{c}} = 6.17 + 3\sqrt{6.17} = 13.62$$
$$LCL \Rightarrow 0$$

MTB > Stat > Control Charts > Attributes Charts > C
To exclude subgroups, select C Chart Options, Estimate tab, Omit 10, 11, 15, 22 from calculations

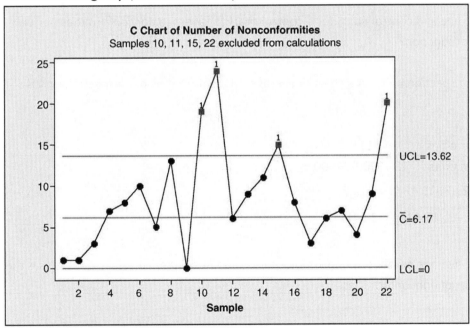

Test Results for C Chart of Ex7.41Num
```
TEST 1. One point more than 3.00 standard deviations from center line.
Test Failed at points:  10, 11, 15, 22
```

For future production, recommend using a *c* chart with center line = 6.17, UCL = 13.62, and LCL = 0.

7.55.

Consider the data in Exercise 7.53. Suppose a new inspection unit is defined as 2,500 m of wire.

(a) What are the center line and control limits for a control chart for monitoring future production based on the total number of nonconformities in the new inspection unit?

The new inspection unit is n = 2500/1000 = 2.5 of the old unit. A c chart of the total number of nonconformities per inspection unit is appropriate.

$$CL = n\bar{c} = 2.5(6.17) = 15.43$$
$$UCL = n\bar{c} + 3\sqrt{n\bar{c}} = 15.43 + 3\sqrt{15.43} = 27.21$$
$$LCL = n\bar{c} - 3\sqrt{n\bar{c}} = 15.43 - 3\sqrt{15.43} = 3.65$$

The plot point, $\hat{c}$, is the total number of nonconformities found while inspecting a sample 2500m in length.

(b) What are the center line and control limits for a control chart for average nonconformities per unit used to monitor future production?

The sample is n =1 new inspection units. A u chart of average nonconformities per inspection unit is appropriate.

$$CL = \bar{u} = \frac{\text{total nonconformities}}{\text{total inspection units}} = \frac{111}{(18 \times 1000)/2500} = 15.42$$
$$UCL = \bar{u} + 3\sqrt{\bar{u}/n} = 15.42 + 3\sqrt{15.42/1} = 27.20$$
$$LCL = \bar{u} - 3\sqrt{\bar{u}/n} = 15.42 - 3\sqrt{15.42/1} = 3.64$$

The plot point, $\hat{u}$, is the average number of nonconformities found in 2500m, and since n = 1, this is the same as the total number of nonconformities.

7.57.

Find the three-sigma control limits for

(a) a c chart with process average equal to four nonconformities.

$$CL = \bar{c} = 4$$
$$UCL = \bar{c} + 3\sqrt{\bar{c}} = 4 + 3\sqrt{4} = 10$$
$$LCL = \bar{c} - 3\sqrt{\bar{c}} = 4 - 3\sqrt{4} \Rightarrow 0$$

(b) a u chart with c = 4 and n = 4.

$$c = 4; \quad n = 4$$
$$CL = \bar{u} = c/n = 4/4 = 1$$
$$UCL = \bar{u} + 3\sqrt{\bar{u}/n} = 1 + 3\sqrt{1/4} = 2.5$$
$$LCL = \bar{u} - 3\sqrt{\bar{u}/n} = 1 - 3\sqrt{1/4} \Rightarrow 0$$

7.59.

Find the three-sigma control limits for

(a) a c chart with process average equal to nine nonconformities.

$$CL = \bar{c} = 9$$
$$UCL = \bar{c} + 3\sqrt{\bar{c}} = 9 + 3\sqrt{9} = 18$$
$$LCL = \bar{c} - 3\sqrt{\bar{c}} = 9 - 3\sqrt{9} = 0$$

(b) a u chart with c = 16 and n = 4.

$$c = 16; \quad n = 4$$
$$CL = \bar{u} = c/n = 16/4 = 4$$
$$UCL = \bar{u} + 3\sqrt{\bar{u}/n} = 4 + 3\sqrt{4/4} = 7$$
$$LCL = \bar{u} - 3\sqrt{\bar{u}/n} = 4 - 3\sqrt{4/4} = 1$$

7.61.

Find 0.975 and 0.025 probability limits for a control chart for nonconformities when $c = 7.6$.

Using the cumulative Poisson distribution:

MTB > Calc > Probability distributions > Poisson
Cumulative Distribution Function
| ```
Poisson with mean = 7.6
  x   P( X <= x )
  2      0.018757
  3      0.055371
 12      0.953566
 13      0.976247
``` |

for the c chart, UCL = 13 and LCL = 2. As a comparison, the normal distribution gives

$$\text{UCL} = \bar{c} + z_{0.975}\sqrt{\bar{c}} = 7.6 + 1.96\sqrt{7.6} = 13.00$$
$$\text{LCL} = \bar{c} - z_{0.025}\sqrt{\bar{c}} = 7.6 - 1.96\sqrt{7.6} = 2.20$$
```

7.63.

The number of workmanship nonconformities observed in the final inspection of disk-drive assemblies has been tabulated as shown in Table 7E.17. Does the process appear to be in control?

■ **TABLE 7E.17**
Data for Exercise 7.63

Day	Number of Assemblies Inspected	Total Number of Imperfections	Day	Number of Assemblies Inspected	Total Number of Imperfections
1	2	10	6	4	24
2	4	30	7	2	15
3	2	18	8	4	26
4	1	10	9	3	21
5	3	20	10	1	8

The appropriate chart is a u chart with control limits based on each sample size:

$$\bar{u} = 7; \quad \text{UCL}_i = 7 + 3\sqrt{7/n_i}; \quad \text{LCL}_i = 7 - 3\sqrt{7/n_i}$$

MTB > Stat > Control Charts > Attributes Charts > U

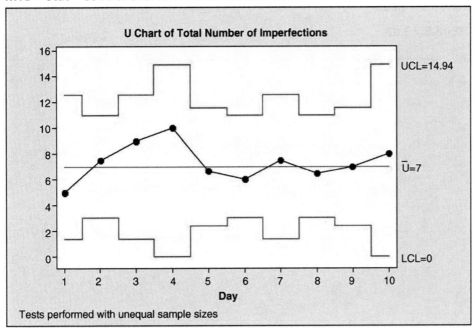

The process is in statistical control.

7.67.

A textile mill wishes to establish a control procedure on flaws in towels it manufactures. Using an inspection unit of 50 units, past inspection data show that 100 previous units had 850 total flaws. What type of control chart is appropriate? Design the control chart such that it has two-sided probability control limits of $\alpha = 0.06$, approximately. Give the center line and control limits.

A c chart with one inspection unit equal to 50 manufacturing units is appropriate. $\bar{c} = 850 / 100 = 8.5$.

MTB > Calc > Probability distributions > Poisson
Cumulative Distribution Function

```
Poisson with mean = 8.5
  x    P( X <= x )
  3      0.030109
 13      0.948589
 14      0.972575
```

LCL = 3 and UCL = 13. For comparison, the normal distribution gives

$$\text{UCL} = \bar{c} + z_{0.97}\sqrt{\bar{c}} = 8.5 + 1.88\sqrt{8.5} = 13.98$$

$$\text{LCL} = \bar{c} + z_{0.03}\sqrt{\bar{c}} = 8.5 - 1.88\sqrt{8.5} = 3.02$$

7.69.

Assembled portable television sets are subjected to a final inspection for surface defects. A total procedure is established based on the requirement that if the average number of nonconformities per units is 4.0, theprobability of concluding that the process is in control will be 0.99. There is to be no lower control limit. What is the appropriate type of control chart and what is the required upper limit?

$\bar{u} = 4.0$ average number of nonconformities/unit. Desire $\alpha = 0.99$. Use the cumulative Poisson distribution to determine the UCL:

```
MTB > Calc > Probability distributions > Poisson
Cumulative Distribution Function

Poisson with mean = 4
 x    P( X <= x )
 0      0.018316
 1      0.091578
 2      0.238103
 3      0.433470
 4      0.628837
 5      0.785130
 6      0.889326
 7      0.948866
 8      0.978637
 9      0.991868
10      0.997160
11      0.999085
```

An UCL = 9 will give a probability of 0.99 of concluding the process is in control, when in fact it is.

7.71.

Consider the situation described in Exercise 7.70. ($\overline{c} = 0.533$)

(a) Find two-sigma control limits and compare these with the control limits found in art (a) of exercise 7.70.

$$\overline{c} \pm 2\sqrt{\overline{c}} = 0.533 + 2\sqrt{0.533} = [0, 1.993]$$

(b) Find the α-risk for the control chart with two-sigma control limits and compare with the results of part (b) of Exercise 7.70.

$$
\begin{aligned}
\alpha &= \Pr\{D < LCL \mid \overline{c}\} + \Pr\{D > UCL \mid \overline{c}\} \\
&= \Pr\{D < 0 \mid 0.533\} + \left[1 - \Pr\{D \leq 1.993 \mid 0.533\}\right] \\
&= 0 + \left[1 - POI(1, 0.533)\right] \\
&= 1 - 0.8996 \\
&= 0.1004
\end{aligned}
$$

where POI($\cdot$) is the cumulative Poisson distribution.

(c) Find the β-risk for $c - 2.0$ for the chart with two-sigma control limits and compare with the results of part (c) of Exercise 7.70.

$$
\begin{aligned}
\beta &= \Pr\{D < UCL \mid c\} - \Pr\{D \leq LCL \mid c\} \\
&= \Pr\{D < 1.993 \mid 2\} - \Pr\{D \leq 0 \mid 2\} \\
&= POI(1, 2) - POI(0, 2) \\
&= 0.406 - 0.135 \\
&= 0.271
\end{aligned}
$$

where POI($\cdot$) is the cumulative Poisson distribution.

(d) Find the ARL if $c = 2.0$ and compare with the ARL found in part (d) of Exercise 7.70.

$$ARL_1 = \frac{1}{1 - \beta} = \frac{1}{1 - 0.271} = 1.372 \approx 2$$

7.73.

A control chart for nonconformities is maintained on a process producing desk calculators. The inspection unit is defined as two calculators. The average number of nonconformities per machine when the process is in control is estimated to be two.

$\bar{u} = $ average # nonconformities/calculator $= 2$

(a) Find the appropriate three-sigma control limits for this size inspection unit.

c chart with $\bar{c} = \bar{u} \times n = 2(2) = 4$ nonconformities/inspection unit

$CL = \bar{c} = 4$

$UCL = \bar{c} + k\sqrt{\bar{c}} = 4 + 3\sqrt{4} = 10$

$LCL = \bar{c} - k\sqrt{\bar{c}} = 4 - 3\sqrt{4} \Rightarrow 0$

(b) What is the probability of type I error for this control chart?

Type I error =

$\alpha = \Pr\{D < LCL \mid \bar{c}\} + \Pr\{D > UCL \mid \bar{c}\}$

$= \Pr\{D < 0 \mid 4\} + [1 - \Pr\{D \le 10 \mid 4\}]$

$= 0 + [1 - POI(10,4)]$

$= 1 - 0.997$

$= 0.003$

where POI($\cdot$) is the cumulative Poisson distribution.

7.75.

Suppose that we wish to design a control chart for nonconformities per unit with L-sigma limits. Find the minimum sample size that would result in a positive lower control limit for this chart.

c: nonconformities per unit; L: sigma control limits

$n\bar{c} - L\sqrt{n\bar{c}} > 0$

$n\bar{c} > L\sqrt{n\bar{c}}$

$n > L^2/\bar{c}$

7.79.

A paper by R.N. Rodriguez ("Health Care Applications of Statistical Process Control: Examples Using the SAS® System," *SAS Users Group International: Proceedings of the 21st Annual Conference*, 1996) illustrated several informative applications of control charts to the health care environment. One of these showed how a control chart was employed to analyze the rate of CAT scans performed each month at a clinic. The data used in this example are shown in Table 7E.21. NSCANB is the number of CAT scans performed each month and MMSB is the number of members enrolled in the health care plan each month, in units of member months. DAYS is the number of days in each month. The variable NYRSB converts MMSB to units of thousand members per year, and is computed as follows: NYRSB = MMSB (Days/30)/12000. NYRSB represents the "area of opportunity." Construct an appropriate control chart to monitor the rate at which CAT scans are performed at this clinic.

▪ TABLE 7E.21
Data for Exercise 7.79

Month	NSCANB	MMSB	Days	NYRSB
Jan. 94	50	26,838	31	2.31105
Feb. 94	44	26,903	28	2.09246
March 94	71	26,895	31	2.31596
Apr. 94	53	26,289	30	2.19075
May 94	53	26,149	31	2.25172
Jun. 94	40	26,185	30	2.18208
July 94	41	26,142	31	2.25112
Aug. 94	57	26,092	31	2.24681
Sept. 94	49	25,958	30	2.16317
Oct. 94	63	25,957	31	2.23519
Nov. 94	64	25,920	30	2.16000
Dec. 94	62	25,907	31	2.23088
Jan. 95	67	26,754	31	2.30382
Feb. 95	58	26,696	28	2.07636
March 95	89	26,565	31	2.28754

The variable NYRSB can be thought of as an "inspection unit", representing an identical "area of opportunity" for each "sample". The "process characteristic" to be controlled is the rate of CAT scans. A *u* chart which monitors the average number of CAT scans per NYRSB is appropriate.

7.79. continued

MTB > Stat > Control Charts > Attributes Charts > U

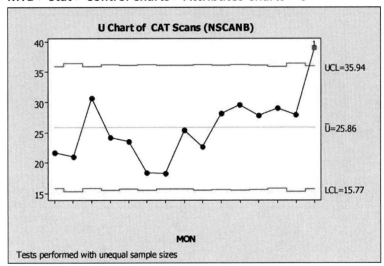

Test Results for U Chart of Ex7.65NSCANB

```
TEST 1. One point more than 3.00 standard deviations from center line.
Test Failed at points:   15
```

The rate of monthly CAT scans is out of control, exceeding the upper control limit in March'95.

Chapter 8

Process and Measurement System Capability Analysis

LEARNING OBJECTIVES

After completing this chapter you should be able to:

1. Investigate and analyze process capability using control charts, histograms, and probability plots
2. Understand the difference between process capability and process potential
3. Calculate and properly interpret process capability ratios
4. Understand the role of the normal distribution interpreting most process capability ratios
5. Calculate confidence intervals on process capability ratios
6. Conduct and analyze a measurement systems capability (or gauge R&R) experiment
7. Estimate the components of variability in a measurement system
8. Set specifications on components in a system involving interaction components to ensure that overall system requirements are met
9. Estimate the natural limits of a process from a sample of data from that process

IMPORTANT TERMS AND CONCEPTS

ANOVA approach to a gauge R&R experiment

Components of gauge error
Components of measurement error
Confidence intervals for gauge R&R studies
Confidence intervals on process capability ratios
Consumer's risk or missed fault for a gauge
Control charts and process capability analysis
Delta method
Discrimination ratio (DR) for a gauge
Estimating variance components
Factorial experiment
Gauge R&R experiment
Graphical methods for process capability analysis
Measurement systems capability analysis
Natural tolerance limits for a normal distribution
Natural tolerance limits of a process
Nonparametric tolerance limits

Normal distribution and process capability ratios
One-sided process capability ratios
PCR C_p
PCR C_{pk}
PCR C_{pm}
Precision versus accuracy of a gauge
Precision-to-tolerance (P/T) ratio
Process capability
Process capability analysis
Process performance indices P_p and P_{pk}
Producer's risk or false failure for a gauge
Product characterization
Random effects model ANOVA
Signal-to-noise ratio (SNR) for a gauge
Tolerance stack-up problems
Transmission of error formula

EXERCISES

8.3.

Consider the piston ring data in Table 6.3. Estimate the process capability assuming that specifications are 74.00 ± 0.035 mm.

$$\hat{\mu} = \bar{\bar{x}} = 74.001; \quad \bar{R} = 0.023; \quad \hat{\sigma} = \bar{R}/d_2 = 0.023/2.326 = 0.010; \quad SL = 74.000 \pm 0.035 = [73.965, 74.035]$$

$$\hat{C}_p \frac{USL - LSL}{6\hat{\sigma}} = \frac{74.035 - 73.965}{6(0.010)} = 1.17$$

$$\hat{C}_{pl} = \frac{\hat{\mu} - LSL}{3\hat{\sigma}} = \frac{74.001 - 73.965}{3(0.010)} = 1.20; \quad \hat{C}_{pu} = \frac{USL - \hat{\mu}}{3\hat{\sigma}} = \frac{74.035 - 74.001}{3(0.010)} = 1.13; \quad \hat{C}_{pk} = \min\left(\hat{C}_{pl}, \hat{C}_{pu}\right) = 1.13$$

Check two important assumptions:

1. Is piston ring inner diameter normally distributed? YES.

MTB > Stat > Basic Statistics > Normality test

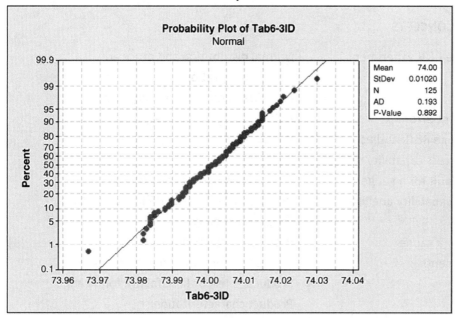

8.3. continued

2. Is the process in statistical control? From $\bar{x}$ and s charts in Example 6.3, YES.

MTB > Stat > Quality Tools > Capability Analysis > Normal
Under Estimate, check that Rbar is used to estimate within subgroup standard deviation

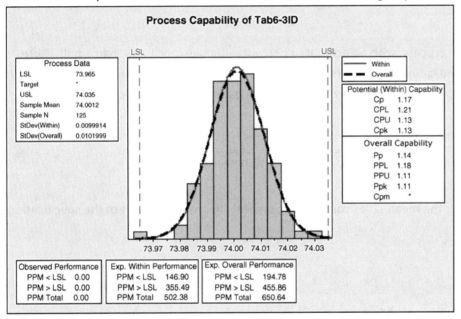

8.5.

Estimate process capability using the $\bar{x}$ and R charts for the power supply voltage data in **Exercise 6.8** (note that early printings of the 7th edition indicate Exercise 6.2). If specifications are at 350 ± 50 V, calculate Cp, Cpk, and Cpkm. Interpret these capability ratios.

$$\hat{\mu} = \bar{\bar{x}} = 10.375; \bar{R}_x = 6.25; \hat{\sigma}_x = \bar{R}/d_2 = 6.25/2.059 = 3.04$$
$$USL_x = [(350+5)-350] \times 10 = 50; LSL_x = [(350-5)-350] \times 10 = -50$$
$$x_i = (obs_i - 350) \times 10$$
$$\hat{C}_p = \frac{USL_x - LSL_x}{6\hat{\sigma}_x} = \frac{50-(-50)}{6(3.04)} = 5.48$$

8.5. continued

The process produces product that uses approximately 18% of the total specification band.

$$\hat{C}_{pu} = \frac{USL_x - \hat{\mu}}{3\hat{\sigma}_x} = \frac{50 - 10.375}{3(3.04)} = 4.34; \quad \hat{C}_{pl} = \frac{\hat{\mu} - LSL_x}{3\hat{\sigma}_x} = \frac{10.375 - (-50)}{3(3.04)} = 6.62$$

$$\hat{C}_{pk} = \min(\hat{C}_{pu}, \hat{C}_{pl}) = 4.34$$

This is an extremely capable process, with an estimated percent defective much less than 1 ppb. Note that the C_{pk} is less than C_p, indicating that the process is not centered and is not achieving potential capability. However, this PCR does not tell *where* the mean is located within the specification band.

$$V = \frac{T - \overline{\overline{x}}}{S} = \frac{0 - 10.375}{3.04} = -3.4128; \quad \hat{C}_{pm} = \frac{\hat{C}_p}{\sqrt{1 + V^2}} = \frac{5.48}{\sqrt{1 + (-3.4128)^2}} = 1.54$$

Since C_{pm} is greater than 4/3, the mean μ lies within approximately the middle fourth of the specification band.

$$\hat{\xi} = \frac{\hat{\mu} - T}{\hat{\sigma}} = \frac{10.375 - 0}{3.04} = 3.41; \quad \hat{C}_{pkm} = \frac{\hat{C}_{pk}}{\sqrt{1 + \hat{\xi}^2}} = \frac{1.54}{\sqrt{1 + 3.41^2}} = 0.43$$

8.7.

A process is in control with $\overline{\overline{x}} = 100, \overline{s} = 1.05$, and $n = 5$. The process specifications are at 95 ± 10. The quality characteristic has a normal distribution.

$$\hat{\mu} = \overline{\overline{x}} = 100; \overline{s} = 1.05; \hat{\sigma}_x = \overline{s}/c_4 = 1.05/0.9400 = 1.117$$

(a) Estimate the potential capability.

$$\hat{C}_p = \frac{USL - LSL}{6\hat{\sigma}} = \frac{(95 + 10) - (95 - 10)}{6(1.117)} = 2.98$$

(b) Estimate the actual capability.

$$\hat{C}_{pl} = \frac{\hat{\mu} - LSL_x}{3\hat{\sigma}_x} = \frac{100 - (95 - 10)}{3(1.117)} = 4.48; \quad \hat{C}_{pu} = \frac{USL_x - \hat{\mu}}{3\hat{\sigma}_x} = \frac{(95 + 10) - 100}{3(1.117)} = 1.49$$

$$\hat{C}_{pk} = \min(\hat{C}_{pl}, \hat{C}_{pu}) = 1.49$$

8.7. continued

(c) How much could the fallout in the process be reduced if the process were corrected to operate at the nominal specification?

$$\hat{p}_{Actual} = Pr\{x < LSL\} + Pr\{x > USL\}$$
$$= Pr\{x < LSL\} + \left[1 - Pr\{x \le USL\}\right]$$
$$= Pr\left\{z < \frac{LSL - \hat{\mu}}{\hat{\sigma}}\right\} + \left[1 - Pr\left\{z \le \frac{USL - \hat{\mu}}{\hat{\sigma}}\right\}\right]$$
$$= Pr\left\{z < \frac{85 - 100}{1.117}\right\} + \left[1 - Pr\left\{z \le \frac{105 - 100}{1.117}\right\}\right]$$
$$= \Phi(-13.429) + \left[1 - \Phi(4.476)\right]$$
$$= 0.0000 + \left[1 - 0.999996\right]$$
$$= 0.000004$$

$$\hat{p}_{Potential} = Pr\left\{z < \frac{85 - 95}{1.117}\right\} + \left[1 - Pr\left\{z \le \frac{105 - 95}{1.117}\right\}\right]$$
$$= \Phi(-8.953) + \left[1 - \Phi(8.953)\right]$$
$$= 0.000000 + \left[1 - 1.000000\right]$$
$$= 0.000000$$

8.9.

A process is in statistical control with $\bar{\bar{x}} = 39.7$ and $\bar{R} = 2.5$. The control chart uses a sample size of $n = 2$. Specifications are at 40 ± 5. The quality characteristic is normally distributed.

$$n = 2; \quad \hat{\mu} = \bar{\bar{x}} = 39.7; \quad \bar{R} = 2.5; \quad \hat{\sigma}_x = \bar{R}/d_2 = 2.5/1.128 = 2.216$$
$$USL = 40 + 5 = 45; \ LSL = 40 - 5 = 35$$

(a) Estimate the potential capability of the process.

$$\hat{C}_p = \frac{USL - LSL}{6\hat{\sigma}} = \frac{45 - 35}{6(2.216)} = 0.75$$

(b) Estimate the actual process capability.

$$\hat{C}_{pu} = \frac{USL - \hat{\mu}}{3\hat{\sigma}} = \frac{45 - 39.7}{3(2.216)} = 0.80; \quad \hat{C}_{pl} = \frac{\hat{\mu} - LSL}{3\hat{\sigma}} = \frac{39.7 - 35}{3(2.216)} = 0.71$$
$$\hat{C}_{pk} = min(\hat{C}_{pl}, \hat{C}_{pu}) = 0.71$$

8.9. continued

(c) Calculate and compare the PCRs Cpm and Cpkm.

$$V = \frac{\bar{x} - T}{s} = \frac{39.7 - 40}{2.216} = -0.135$$

$$\hat{C}_{pm} = \frac{\hat{C}_p}{\sqrt{1 + V^2}} = \frac{0.75}{\sqrt{1 + (-0.135)^2}} = 0.74$$

$$\hat{C}_{pkm} = \frac{\hat{C}_{pk}}{\sqrt{1 + V^2}} = \frac{0.71}{\sqrt{1 + (-0.135)^2}} = 0.70$$

The closeness of estimates for C_p, C_{pk}, C_{pm}, and C_{pkm} indicate that the process mean is very close to the specification target.

(d) How much improvement could be made in process performance if the mean could be centered at the nominal value?

The current fraction nonconforming is:

$$\hat{p}_{Actual} = \Pr\{x < LSL\} + \Pr\{x > USL\}$$
$$= \Pr\{x < LSL\} + [1 - \Pr\{x \le USL\}]$$
$$= \Pr\left\{z < \frac{LSL - \hat{\mu}}{\hat{\sigma}}\right\} + \left[1 - \Pr\left\{z \le \frac{USL - \hat{\mu}}{\hat{\sigma}}\right\}\right]$$
$$= \Pr\left\{z < \frac{35 - 39.7}{2.216}\right\} + \left[1 - \Pr\left\{z \le \frac{45 - 39.7}{2.216}\right\}\right]$$
$$= \Phi(-2.12094) + [1 - \Phi(2.39170)]$$
$$= 0.0169634 + [1 - 0.991615]$$
$$= 0.025348$$

If the process mean could be centered at the specification target, the fraction nonconforming would be:

$$\hat{p}_{Potential} = 2 \times \Pr\left\{z < \frac{35 - 40}{2.216}\right\}$$
$$= 2 \times \Pr\{z < -2.26\}$$
$$= 2 \times 0.01191$$
$$= 0.02382$$

8.11.

Consider the two processes shown in Table 8E.1 (the sample size $n = 5$).

■ TABLE 8E.1
Process Data for Exercise 8.9

Process A	Process B
$\bar{\bar{x}}_A = 100$	$\bar{\bar{x}}_B = 105$
$\bar{s}_A = 3$	$\bar{s}_B = 1$

Specifications are at 100 ± 10. Calculate Cp, Cpk, and Cpm and interpret these ratios. Which process would you prefer to use?

<u>Process A</u>

$$\hat{\mu} = \bar{\bar{x}}_A = 100; \bar{s}_A = 3; \hat{\sigma}_A = \bar{s}_A/c_4 = 3/0.9400 = 3.191$$

$$\hat{C}_p = \frac{USL - LSL}{6\hat{\sigma}} = \frac{(100+10)-(100-10)}{6(3.191)} = 1.045$$

$$\hat{C}_{pu} = \frac{USL_x - \hat{\mu}}{3\hat{\sigma}_x} = \frac{(100+10)-100}{3(3.191)} = 1.045; \quad \hat{C}_{pl} = \frac{\hat{\mu} - LSL_x}{3\hat{\sigma}_x} = \frac{100-(100-10)}{3(3.191)} = 1.045$$

$$\hat{C}_{pk} = \min(\hat{C}_{pl}, \hat{C}_{pu}) = 1.045$$

$$V = \frac{\bar{x} - T}{s} = \frac{100-100}{3.191} = 0; \quad \hat{C}_{pm} = \frac{\hat{C}_p}{\sqrt{1+V^2}} = \frac{1.045}{\sqrt{1+(0)^2}} = 1.045$$

$$\hat{p} = \Pr\{x < LSL\} + \Pr\{x > USL\}$$
$$= \Pr\{x < LSL\} + 1 - \Pr\{x \le USL\}$$
$$= \Pr\left\{z < \frac{LSL - \hat{\mu}}{\hat{\sigma}}\right\} + 1 - \Pr\left\{z \le \frac{USL - \hat{\mu}}{\hat{\sigma}}\right\}$$
$$= \Pr\left\{z < \frac{90-100}{3.191}\right\} + 1 - \Pr\left\{z \le \frac{110-100}{3.191}\right\}$$
$$= \Phi(-3.13) + 1 - \Phi(3.13)$$
$$= 0.00087 + 1 - 0.99913$$
$$= 0.00174$$

8.11. continued

<u>Process B</u>

$$\hat{\mu} = \overline{\overline{x}}_B = 105; \; \overline{s}_B = 1; \; \hat{\sigma}_B = \overline{s}_B / c_4 = 1/0.9400 = 1.064$$

$$\hat{C}_p = \frac{USL - LSL}{6\hat{\sigma}} = \frac{(100+10)-(100-10)}{6(1.064)} = 3.133$$

$$\hat{C}_{pl} = \frac{\hat{\mu}_x - LSL_x}{3\hat{\sigma}_x} = \frac{105-(100-10)}{3(1.064)} = 4.699; \quad \hat{C}_{pu} = \frac{USL_x - \hat{\mu}_x}{3\hat{\sigma}_x} = \frac{(100+10)-105}{3(1.064)} = 1.566$$

$$\hat{C}_{pk} = \min(\hat{C}_{pl}, \hat{C}_{pu}) = 1.566$$

$$V = \frac{\overline{x} - T}{s} = \frac{100-105}{1.064} = -4.699; \quad \hat{C}_{pm} = \frac{\hat{C}_p}{\sqrt{1+V^2}} = \frac{3.133}{\sqrt{1+(-4.699)^2}} = 0.652$$

$$\hat{p} = \Pr\left\{ z < \frac{90-105}{1.064} \right\} + 1 - \Pr\left\{ z \le \frac{110-105}{1.064} \right\}$$

$$= \Phi(-14.098) + 1 - \Phi(4.699)$$

$$= 0.000000 + 1 - 0.999999$$

$$= 0.000001$$

Prefer to use Process B with estimated process fallout of 0.000001 instead of Process A with estimated fallout 0.001726.

8.13.

The weights of nominal 1-kg containers of a concentrated chemical ingredient are shown in Table 8E.2. Prepare a normal probability plot of the data and estimate process capability. Does this conclusion depend on process stability?

▪ **TABLE 8E.2**
Weights of Containers

0.9475	0.9775	0.9965	1.0075	1.0180
0.9705	0.9860	0.9975	1.0100	1.0200
0.9770	0.9960	1.0050	1.0175	1.0250

MTB > Graph > Probability Plot > Single

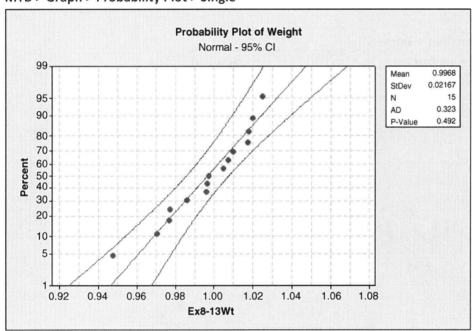

A normal probability plot of the 1-kg container weights shows the distribution is close to normal.

$$\bar{x} \approx p_{50} = 0.9975; \quad p_{84} = 1.0200$$
$$\hat{\sigma} = p_{84} - p_{50} = 1.0200 - 0.9975 = 0.0225$$
$$6\hat{\sigma} = 6(0.0225) = 0.1350$$

8.17.

The height of the disk used in a computer disk drive assembly is a critical quality characteristic. Table 8E.5 gives the heights (in mm) of 25 disks randomly selected from the manufacturing process. Assume that the process is in statistical control. Prepare a normal probability plot of the disk height data and estimate process capability.

■ TABLE 8E.5
Disk Height Data for Exercise 8.13

20.0106	20.0090	20.0067	19.9772	20.0001
19.9940	19.9876	20.0042	19.9986	19.9958
20.0075	20.0018	20.0059	19.9975	20.0089
20.0045	19.9891	19.9956	19.9884	20.0154
20.0056	19.9831	20.0040	20.0006	20.0047

MTB > Graph > Probability Plot > Single
Select Scale, then the tab for Percentile lines to add lines at Y values 50 and 84

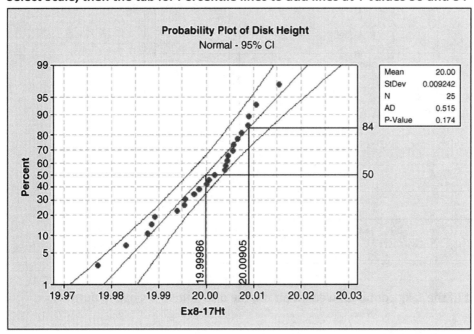

A normal probability plot of computer disk heights shows the distribution is close to normal.

$$\bar{x} \approx p_{50} = 19.99986$$
$$p_{84} = 20.00905$$
$$\hat{\sigma} = p_{84} - p_{50} = 20.00905 - 19.99986 = 0.00919$$
$$6\hat{\sigma} = 6(0.00919) = 0.05514$$

8.19.

An electric utility tracks the response time to customer-reported outages. The data in Table 8E.7 are a random sample of 40 of the response times (in minutes) for one operating division of this utility during a single month.

■ TABLE 8E.7
Response Time Data for Exercise 8.19

80	102	86	94	86	106	105	110	127	97
110	104	97	128	98	84	97	87	99	94
105	104	84	77	125	85	80	104	103	109
115	89	100	96	96	87	106	100	102	93

(a) Estimate the capability of the utility's process for responding to customer-reported outages.

MTB > Graph > Probability Plot > Single
Select Scale, then the tab for Percentile lines to add lines at Y values 50 and 84

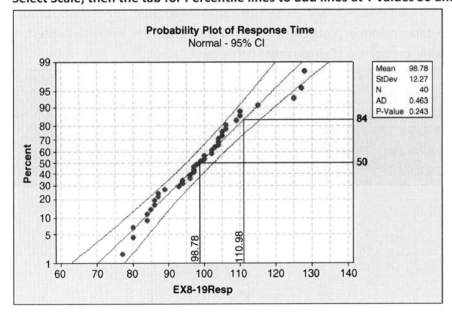

A normal probability plot of response times shows the distribution is close to normal.

$$\bar{x} \approx p_{50} = 98.78; \quad p_{84} = 110.98; \quad \hat{\sigma} = p_{84} - p_{50} = 110.98 - 98.78 = 12.2; \quad 6\hat{\sigma} = 6(12.2) = 73.2$$

8.19. continued

(b) The utility wants to achieve a 90% response rate in under two hours, as a response to emergency outages is an important measure of customer satisfaction. What is the capability of the process with respect to this objective.

USL = 2 hrs = 120 mins

$$C_{pu} = \frac{USL - \hat{\mu}}{3\hat{\sigma}} = \frac{120 - 98.78}{3(12.2)} = 0.58$$

$$\hat{p} = \Pr\left\{z > \frac{USL - \hat{\mu}}{\hat{\sigma}}\right\} = 1 - \Pr\left\{z < \frac{USL - \hat{\mu}}{\hat{\sigma}}\right\} = 1 - \Pr\left\{z < \frac{120 - 98.78}{12.2}\right\}$$

$$= 1 - \Phi(1.739) = 1 - 0.958983 = 0.041017$$

8.21.

The failure time in hours of ten LSI memory devices follows: 1210, 1275, 1400, 1695, 1900, 2105, 2230, 2250, 2500, and 2625. Plot the data on normal probability paper and, if appropriate, estimate process capability. Is it safe to estimate the proportion of circuits that fail below 1,200h?

MTB > Graph > Probability Plot > Single
Select Scale, then the tab for Percentile lines to add lines at Y values 50 and 84

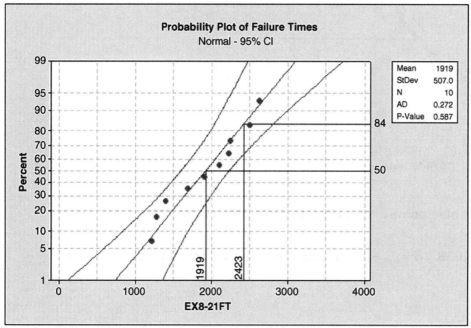

8.21. continued

The data lie nearly along a straight line, implying that the distribution is normal, and it is appropriate to estimate process capability and the proportion of circuits below the LSL.

$\bar{x} \approx P_{50} = 1919; \quad P_{84} = 2423$

$\hat{\sigma} = P_{84} - P_{50} = 2423 - 1919 = 504$

$p = \Pr\{x < 1200\} = \Phi\left(\dfrac{LSL - \hat{\mu}}{\hat{\sigma}}\right) = \Phi\left(\dfrac{1200 - 1919}{504}\right) = \Phi(1.42) = 0.07685$

8.23.

A company has been asked by an important customer to demonstrate that its process capability ratio Cp exceeds 1.33. It has taken a sample of 50 parts and obtained the point estimate $\hat{C}_p = 1.52$. Assume that the quality characteristic follows a normal distribution. Can the company demonstrate that Cp exceeds 1.33 at the 95% level of confidence? What level of confidence would give a one-sided lower confidence limit on Cp that exceeds 1.33?

$1 - \alpha = 0.95; \quad \chi^2_{1-\alpha, n-1} = \chi^2_{0.95, 49} = 33.9303$

$\hat{C}_p \sqrt{\dfrac{\chi^2_{1-\alpha, n-1}}{n-1}} \leq C_p$

$1.52 \sqrt{\dfrac{33.9303}{49}} = 1.26 \leq C_p$

The company cannot demonstrate that the PCR exceeds 1.33 at a 95% confidence level.

$1.52 \sqrt{\dfrac{\chi^2_{1-\alpha, 49}}{49}} = 1.33$

$\chi^2_{1-\alpha, 49} = 49\left(\dfrac{1.33}{1.52}\right)^2 = 37.52$

$1 - \alpha = 0.88$

$\alpha = 0.12$

8.25.

The molecular weight of a particular polymer should fall between 2,100 and 2,350. Fifty samples of this material were analyzed with the results $\bar{x} = 2,275$ and $s = 60$. Assume that molecular weight is normally distributed.

(a) Calculate a point estimate of Cpk.

$$\hat{C}_{pu} = \frac{USL_x - \hat{\mu}_x}{3\hat{\sigma}_x} = \frac{2350 - 2275}{3(60)} = 0.42$$

$$\hat{C}_{pl} = \frac{\hat{\mu}_x - LSL_x}{3\hat{\sigma}_x} = \frac{2275 - 2100}{3(60)} = 0.97$$

$$\hat{C}_{pk} = \min(\hat{C}_{pl}, \hat{C}_{pu}) = 0.42$$

(b) Find a 95% confidence interval on Cpk.

$$\alpha = 0.05; z_{\alpha/2} = z_{0.025} = 1.960$$

$$\hat{C}_{pk}\left[1 - z_{\alpha/2}\sqrt{\frac{1}{9n\hat{C}_{pk}^2} + \frac{1}{2(n-1)}}\right] \leq C_{pk} \leq \hat{C}_{pk}\left[1 + z_{\alpha/2}\sqrt{\frac{1}{9n\hat{C}_{pk}^2} + \frac{1}{2(n-1)}}\right]$$

$$0.42\left[1 - 1.96\sqrt{\frac{1}{9(50)(0.42)^2} + \frac{1}{2(50-1)}}\right] \leq C_{pk} \leq 0.42\left[1 + 1.96\sqrt{\frac{1}{9(50)(0.42)^2} + \frac{1}{2(50-1)}}\right]$$

$$0.30 \leq C_{pk} \leq 0.54$$

8.29.

An operator-instrument combination is known to test parts with an average error of zero; however, the standard deviation of measurement error is estimated to be 3. Samples from a controlled process were analyzed, and the total variability was estimated to be $\hat{\sigma} = 5$. What is the true process standard deviation?

$$\sigma_{OI} = 0; \hat{\sigma}_I = 3; \hat{\sigma}_{Total} = 5$$

$$\hat{\sigma}_{Total}^2 = \hat{\sigma}_{Meas}^2 + \hat{\sigma}_{Process}^2$$

$$\hat{\sigma}_{Process} = \sqrt{\hat{\sigma}_{Total}^2 - \hat{\sigma}_{Meas}^2} = \sqrt{5^2 - 3^2} = 4$$

8.31.

Ten parts are measured three times by the same operator in a gauge capability study. The data are shown in Table 8E.9

■ **TABLE 8E.9**
Measurement Data for Exercise 8.31

Part Number	Measurements		
	1	2	3
1	100	101	100
2	95	93	97
3	101	103	100
4	96	95	97
5	98	98	96
6	99	98	98
7	95	97	98
8	100	99	98
9	100	100	97
10	100	98	99

(a) Describe the measurement error that results from the use of this gauge.

MTB > Stat > Control Charts > Variables Charts for Subgroups > X-bar R

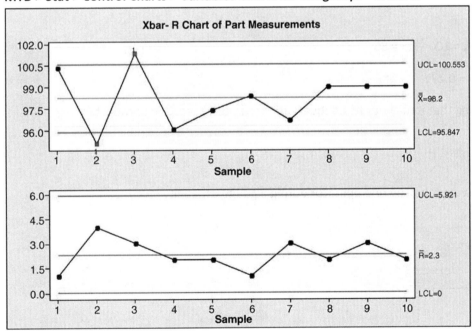

Test Results for Xbar Chart of Ex8.25All

```
TEST 1. One point more than 3.00 standard deviations from center line.
Test Failed at points:  2, 3
```

8.31.(a) continued

The $\bar{x}$ chart has a couple out-of-control points, and the R chart is in control. This indicates that the operator is not having difficulty making consistent measurements.

(b) Estimate the total variability and product variability.

$$\bar{\bar{x}} = 98.2; \bar{R} = 2.3; \hat{\sigma}_{Gauge} = \bar{R}/d_2 = 2.3/1.693 = 1.359$$

$$\hat{\sigma}^2_{Total} = 4.717$$

$$\hat{\sigma}^2_{Product} = \hat{\sigma}^2_{Total} - \hat{\sigma}^2_{Gauge} = 4.717 - 1.359^2 = 2.872$$

$$\hat{\sigma}_{Product} = 1.695$$

(c) What percentage of total variability is due to the gauge?

$$\frac{\hat{\sigma}_{Gauge}}{\hat{\sigma}_{Total}} \times 100 = \frac{1.359}{\sqrt{4.717}} \times 100 = 62.5\%$$

(d) If specifications on the part are at 100 ± 15, find the P/T ratio for this gauge. Comment on the adequacy of the gauge.

USL = 100 + 15 = 115; LSL = 100 − 15 = 85

$$\frac{P}{T} = \frac{6\hat{\sigma}_{Gauge}}{USL - LSL} = \frac{6(1.359)}{115 - 85} = 0.272$$

P/T exceeds 0.1, indicating that it may not be adequate for accurately measuring product.

8.33.

The data in Table 8E.11 were taken by one operator during a gauge capability study.

■ **TABLE 8E.11**
Measurement Data for Exercise 8.33

Part Number	Measurements 1	2	Part Number	Measurements 1	2
1	20	20	9	20	20
2	19	20	10	23	22
3	21	21	11	28	22
4	24	20	12	19	25
5	21	21	13	21	20
6	25	26	14	20	21
7	18	17	15	18	18
8	16	15			

8.33. continued

(a) Estimate gauge capability.

$$\hat{\sigma}_{Gauge} = \overline{R}/d_2 = 1.533/1.128 = 1.359$$

Gauge capability: $6\hat{\sigma} = 8.154$

(b) Does the control chart analysis of these data indicate any potential problem in using the gauge?

MTB > Stat > Control Charts > Variables Charts for Subgroups > X-bar R

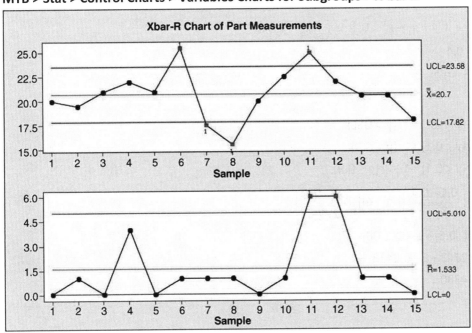

Test Results for R Chart of Ex8.27All

```
TEST 1. One point more than 3.00 standard deviations from center line.
Test Failed at points:  11, 12
```

Out-of-control points on *R* chart indicate operator difficulty with using gage.

8.37.

Two parts are assembled as shown in the figure. The distributions of $x1$ and $x2$ are normal, with $\mu_1 = 20, \sigma_1 = 0.3, \mu_2 = 19.6, \sigma_2 = 0.4$. The specifications of the clearance between the mating parts are 0.5 ± 0.4. What fraction of assemblies will fail to meet specifications if assembly is at random?

$x_1 \sim N(20, 0.3^2); x_2 \sim N(19.6, 0.4^2)$

Nonconformities will occur if $y = x_1 - x_2 < 0.1$ or $y = x_1 - x_2 > 0.9$

$\mu_y = \mu_1 - \mu_2 = 20 - 19.6 = 0.4$

$\sigma_y^2 = \sigma_1^2 + \sigma_2^2 = 0.3^2 + 0.4^2 = 0.25$

$\sigma_y = 0.50$

$$\begin{aligned}
Pr\{Nonconformities\} &= Pr\{y < LSL\} + Pr\{y > USL\} \\
&= Pr\{y < 0.1\} + Pr\{y > 0.9\} \\
&= Pr\{y < 0.1\} + 1 - Pr\{y \le 0.9\} \\
&= \Phi\left(\frac{0.1 - 0.4}{\sqrt{0.25}}\right) + 1 - \Phi\left(\frac{0.9 - 0.4}{\sqrt{0.25}}\right) \\
&= \Phi(-0.6) + 1 - \Phi(1.00) \\
&= 0.2743 + 1 - 0.8413 \\
&= 0.4330
\end{aligned}$$

8.39.

A rectangular piece of metal of width W and length L is cut from a plate of thickness T. If W, L, and T are independent random variables with means and standard deviations as given here and the density of the metal is 0.08 g/cm^3, what would be the estimated mean and standard deviation of the weights of pieces produced by this process?

Variable	Mean	Standard Deviation
W	10 cm	0.2 cm
L	20 cm	0.3 cm
T	3 cm	0.1 cm

8.39. continued

$Weight = d \times W \times L \times T$

$$\cong d\left[\mu_W \mu_L \mu_T + (W - \mu_W)\mu_L \mu_T + (L - \mu_L)\mu_W \mu_T + (T - \mu_T)\mu_W \mu_L\right]$$

$$\hat{\mu}_{Weight} \cong d[\mu_W \mu_L \mu_T] = 0.08(10)(20)(3) = 48$$

$$\hat{\sigma}^2_{Weight} \cong d^2\left[\hat{\mu}^2_W \hat{\sigma}^2_L \hat{\sigma}^2_T + \hat{\mu}^2_L \hat{\sigma}^2_W \hat{\sigma}^2_T + \hat{\mu}^2_T \hat{\sigma}^2_W \hat{\sigma}^2_L\right]$$

$$= 0.08^2\left[10^2(0.3^2)(0.1^2) + 20^2(0.2^2)(0.1^2) + 3^2(0.2^2)(0.3^2)\right] = 0.00181$$

$$\hat{\sigma}_{Weight} \cong 0.04252$$

8.41.

Two resistors are connected to a battery as shown in the figure. Find approximate expressions for the mean and variance of the resulting current (*I*). *E*, R_1, and R_2 are random variables with means $\mu_E, \mu_{R1}, \mu_{R2}$ and variances $\sigma^2_E, \sigma^2_{R1}, \sigma^2_{R2}$, respectively.

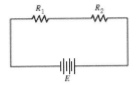

$$I = E/(R_1 + R_2)$$

$$\mu_I \cong \mu_E/(\mu_{R_1} + \mu_{R_2})$$

$$\sigma^2_I \cong \frac{\sigma^2_E}{(\mu_{R_1} + \mu_{R_2})} + \frac{\mu_E}{(\mu_{R_1} + \mu_{R_2})^2}\left(\sigma^2_{R_1} + \sigma^2_{R_2}\right)$$

8.43.

An assembly of two parts is formed by fitting a shaft into a bearing. It is known that the inside diameters of bearings are normally distributed with mean 2.010 cm and standard deviation 0.002 cm, and that the outside diameters of the shafts are normally distributed with mean 2.004 cm and standard deviation 0.001 cm. Determine the distribution of clearance between the parts if random assembly is used. What is the probability that the clearance is positive?

8.43. continued

$ID \sim N(2.010, 0.002^2)$ and $OD \sim N(2.004, 0.001^2)$

Interference occurs if $y = ID - OD < 0$

$\mu_y = \mu_{ID} - \mu_{OD} = 2.010 - 2.004 = 0.006$

$\sigma_y^2 = \sigma_{ID}^2 + \sigma_{OD}^2 = 0.002^2 + 0.001^2 = 0.000005$

$\sigma_y = 0.002236$

$$
\begin{aligned}
Pr\{\text{positive clearance}\} &= 1 - Pr\{\text{interference}\} \\
&= 1 - Pr\{y < 0\} \\
&= 1 - \Phi\left(\frac{0 - 0.006}{\sqrt{0.000005}}\right) \\
&= 1 - \Phi(-2.683) \\
&= 1 - 0.0036 \\
&= 0.9964
\end{aligned}
$$

8.45.

A sample of ten items from a normal population had a mean of 300 and standard deviation of 10. Using these data, estimate a value for the random variable such that the probability is 0.95 that 90% of the measurements on this random variable will lie below the value.

$n = 10$; $x \sim N(300, 10^2)$; $\alpha = 0.10$; $\gamma = 0.95$; one-sided

From Appendix VIII: $K = 2.355$

$UTL = \bar{x} + KS = 300 + 2.355(10) = 323.55$

8.47.

A sample of 20 measurements on a normally distributed quality characteristic had $\bar{x} = 350$ and $s = 10$. Find an upper natural tolerance limit that has probability 0.90 of containing 95% of the distribution of this quality characteristic.

$n = 20$; $x \sim N(350, 10^2)$; $\alpha = 0.05$; $\gamma = 0.90$; one-sided

From Appendix VIII: $K = 2.208$

$UTL = \bar{x} + KS = 350 + 2.208(10) = 372.08$

8.49.

A random sample of $n = 40$ pipe sections resulted in a mean wall thickness of 0.1264 in. and a standard deviation of 0.0003 in. We assume that wall thickness is normally distributed.

(a) Between what limits can we say with 95% confidence that 95% of the wall thicknesses should fall?

$\alpha = 0.05$; $\gamma = 0.95$; and two-sided
From Appendix VII: $K = 2.445$
TI on x: $\bar{x} \pm KS = 0.1264 \pm 2.445(0.0003) = [0.1257, 0.1271]$

(b) Construct a 95% confidence interval on the true mean thickness. Explain the difference between this interval and the one constructed in part (a).

$\alpha = 0.05; t_{\alpha/2, n-1} = t_{0.025, 39} = 2.023$
CI on $\bar{x}$: $\bar{x} \pm t_{\alpha/2, n-1} S/\sqrt{n} = 0.1264 \pm 2.023\left(0.0003/\sqrt{40}\right) = [0.1263, 0.1265]$

Part (a) is a tolerance interval on individual thickness observations; part (b) is a confidence interval on mean thickness. In part (a), the interval relates to individual observations (random variables), while in part (b) the interval refers to a parameter of a distribution (an unknown constant).

Chapter 9

Cumulative Sum and Exponentially Weighted Moving Average Control Charts

LEARNING OBJECTIVES

After completing this chapter you should be able to:

1. Set up and use CUSUM control charts for monitoring the process mean
2. Design a CUSUM control chart for the mean to obtain specific ARL performance
3. Incorporate a fast initial response feature into the CUSUM control chart
4. Use a combined Shewhart-CUSUM monitoring scheme
5. Set up and use EWMA control charts for monitoring the process mean
6. Design an EWMA control chart for the mean to obtain specific ARL performance
7. Understand why the EWMA control chart is robust to the assumption of normality
8. Understand the performance advantage of CUSUM and EWMA control charts relative to Shewhart control charts
9. Set up and use a control chart based on ordinary (unweighted) moving average

IMPORTANT TERMS AND CONCEPTS

ARL calculations for the CUSUM
Average run length
Combined CUSUM-Shewhart procedures
CUSUM control chart
CUSUM status chart
Decision interval
Design of a CUSUM
Design of an EWMA control chart
EWMA control chart
Fast initial response (FIR) or headstart feature for a CUSUM
Fast initial response (FIR) or headstart feature for an EWMA

Moving average control chart
One-sided CUSUMs
Poisson EWMA
Reference value
Robustness of the EWMA to normality
Scale CUSUM
Self-starting CUSUM
Signal resistance of a control chart
Standardized CUSUM
Tabular or algorithmic CUSUM

V-mask form of the CUSUM

Exercises

9.1.

The data in Table 9E.1 represent individual observations on molecular weight taken hourly from a chemical process. The target value of molecular weight is 1,050 and the process standard deviation is thought to be about $\sigma = 25$.

■ TABLE 9E.1
Molecular Weight

Observation Number	x	Observation Number	x
1	1,045	11	1,139
2	1,055	12	1,169
3	1,037	13	1,151
4	1,064	14	1,128
5	1,095	15	1,238
6	1,008	16	1,125
7	1,050	17	1,163
8	1,087	18	1,188
9	1,125	19	1,146
10	1,146	20	1,167

(a) Set up a tabular CUSUM for the mean of this process. Design the CUSUM to quickly detect a shift of about 1.0σ in the process mean.

$\mu_0 = 1050$; $\sigma = 25$; $\delta = 1\sigma$, $K = (\delta/2)\sigma = (1/2)25 = 12.5$; $H = 5\sigma = 5(25) = 125$

9.1.(a) continued

MTB > Stat > Control Charts > Time-Weighted Charts > CUSUM
Enter Subgroup size = 1 and Target = 1050
In CUSUM Options, enter Standard deviation = 25, and on the Plan/Type tab, select One-sided type of
CUSUM, and set h = 5 and k = 0.5

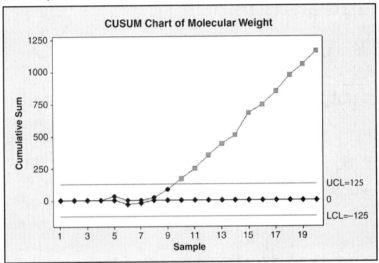

Test Results for CUSUM Chart of Ex9-1mole
```
TEST. One point beyond control limits.
Test Failed at points:   10, 11, 12, 13, 14, 15, 16, 17, 18, 19, 20
```

The process signals out of control at observation 10. The point at which the assignable cause occurred
can be determined by counting the number of increasing plot points. The assignable cause occurred
after observation 10 − 3 = 7.

(b) Is the estimate of σ used in part (a) of this problem reasonable?

$$\hat{\sigma} = \overline{MR2}/d_2 = 38.8421 / 1.128 = 34.4345$$

No. The estimate used for σ is much smaller than that from the data.

9.3.

(a) Add a headstart feature to the CUSUM in Exercise 9.1.

μ_0 = 1050, σ = 25, k = 0.5, K = 12.5, h = 5, $H/2$ = 125/2 = 62.5
FIR = $H/2$ = 62.5, in std dev units = 62.5/25 = 2.5

MTB > Stat > Control Charts > Time-Weighted Charts > CUSUM

Enter Subgroup size = 1 and Target = 0

In CUSUM Options, enter Standard deviation = 1; and on the Plan/Type tab, select One-sided type of CUSUM, enter FIR = 2.5, and set h = 5 and k = 0.5

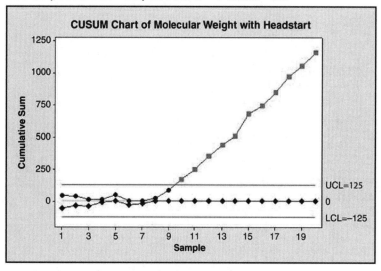

Test Results for CUSUM Chart of Ex9-1mole

```
TEST. One point beyond control limits.
Test Failed at points:   10, 11, 12, 13, 14, 15, 16, 17, 18, 19, 20
```

For example,

$$C_1^+ = \max\left[0, x_i - (\mu_0 - K) + C_0^+\right] = \max\left[0, 1045 - (1050 + 12.5) + 62.5\right] = 45$$

Using the tabular CUSUM, the process signals out of control at observation 10, the same as the CUSUM without a FIR feature.

9.3. continued

(b) Use a combined Shewhart-CUSUM scheme on the data in Exercise 9.1. Interpret the results of both charts.

MTB > Stat > Control Charts > Variables Charts for Individuals > I-MR
In I-MR Options, select S limits tab and enter 3.5 for multiple of standard deviation units

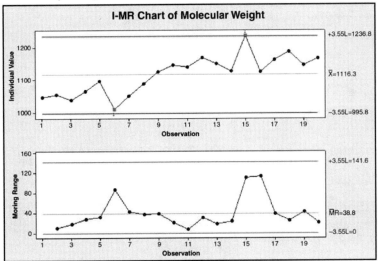

Using 3.5σ limits on the Individuals chart, there are no out-of-control signals. However there does appear to be a trend up from observations 6 through 12—this is the situation detected by the cumulative sum.

9.5.

Rework Exercise 9.4 using the standardized CUSUM parameters of $h = 8.01$ and $k = 0.25$. Compare the results with those obtained previously in Exercise 9.4. What can you say about the theoretical performance of those two CUSUM schemes?

To standardize, let $y_i = \dfrac{x_i - \mu_0}{\sigma} = \dfrac{x_i - 8.02}{0.05}$

9.5. continued

MTB > Stat > Control Charts > Time-Weighted Charts > CUSUM
Enter Subgroup size = 1 and Target = 0
In CUSUM Options, enter Standard deviation = 1, and on the Plan/Type tab, select One-sided type of CUSUM, and set h = 8.01 and k = 0.25

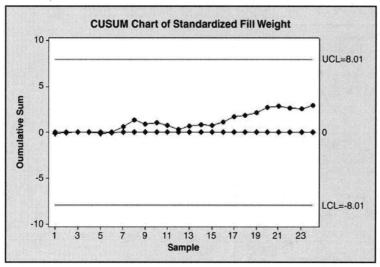

There are no out-of-control signals.

What can you say about the theoretical performance of those two CUSUM schemes?

<u>In Exercise 9.5:</u>

$\mu_0 = 8.02; \sigma = 0.05; k = 0.25; h = 8.01; b = h + 1.166 = 8.01 + 1.166 = 9.176$

$\delta^* = 0; \quad \Delta^+ = \delta^* - k = 0 - 0.25 = -0.25; \quad \Delta^- = -\delta^* - k = -0 - 0.25 = -0.25$

$ARL_0^+ = ARL_0^- \cong \dfrac{\exp[-2(-0.25)(9.176)] + 2(-0.25)(9.176) - 1}{2(-0.25)^2} = 741.6771$

$\dfrac{1}{ARL_0} = \dfrac{1}{ARL_0^+} + \dfrac{1}{ARL_0^-} = \dfrac{2}{741.6771} = 0.0027$

$ARL_0 = 1/0.0027 = 370.84$

9.5. continued

<u>In Exercise 9.4:</u>

$\mu_0 = 8.02; \sigma = 0.05; k = 1/2; h = 4.77; b = h + 1.166 = 4.77 + 1.166 = 5.936$

$\delta^* = 0; \quad \Delta^+ = \delta^* - k = 0 - 0.5 = -0.5; \quad \Delta^- = -\delta^* - k = -0 - 0.5 = -0.5$

$$ARL_0^+ = ARL_0^- \cong \frac{\exp[-2(-0.5)(5.936)] + 2(-0.5)(5.936) - 1}{2(-0.5)^2} = 742.964$$

$$\frac{1}{ARL_0} = \frac{1}{ARL_0^+} + \frac{1}{ARL_0^-} = \frac{2}{742.964} = 0.0027$$

$$ARL_0 = 1/0.0027 = 371.48$$

The theoretical performance of the two CUSUM schemes in Exercises 9.4 and 9.5 is very similar.

9.7.

The data in Table 9E.3 are temperature readings from a chemical process in °C, taken every two minutes. (Read the observations down, from left.) The target value for the mean is $\mu_0 = 950$.

■ **TABLE 9E.3**
Chemical Process Temperature Data

953	985	949	937	959	948	958	952
945	973	941	946	939	937	955	931
972	955	966	954	948	955	947	928
945	950	966	935	958	927	941	937
975	948	934	941	963	940	938	950
970	957	937	933	973	962	945	970
959	940	946	960	949	963	963	933
973	933	952	968	942	943	967	960
940	965	935	959	965	950	969	934
936	973	941	956	962	938	981	927

(a) Estimate the process standard deviation.

$\hat{\sigma} = \overline{MR2}/d_2 = 13.7215/1.128 = 12.16$ (from a Moving Range chart with CL = 13.7215)

9.7. continued

(b) Set up and apply a tabular CUSUM for this process, using standardized values $h = 5$ and $k = \frac{1}{2}$. Interpret this chart.

MTB > Stat > Control Charts > Time-Weighted Charts > CUSUM
Enter Subgroup size = 1 and Target = 950
In CUSUM Options, enter Standard deviation = 12.16, and on the Plan/Type tab, select One-sided type of CUSUM, and set h = 5 and k = 0.5

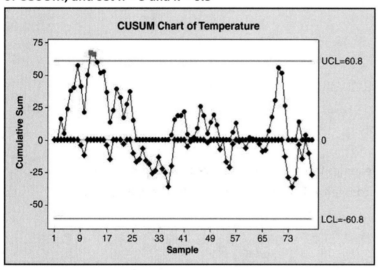

Test Results for CUSUM Chart of Ex9-7temp

```
TEST. One point beyond control limits.
Test Failed at points:  12, 13
```

The process signals out of control at observation 12. The assignable cause occurred after observation $12 - 10 = 2$.

9.9.

Viscosity measurements on a polymer are made every 10 minutes by an on-line viscometer. Thirty-six observations are shown in Table 9E.5 (read down from left). The target viscosity for this process is $\mu_0 = 3{,}200$.

■ **TABLE 9E.5**
Polymer Viscosity

3,169	3,205	3,185	3,188
3,173	3,203	3,187	3,183
3,162	3,209	3,192	3,175
3,154	3,208	3,199	3,174
3,139	3,211	3,197	3,171
3,145	3,214	3,193	3,180
3,160	3,215	3,190	3,179
3,172	3,209	3,183	3,175
3,175	3,203	3,197	3,174

9.9. continued

(a) Estimate the process standard deviation.

$\hat{\sigma} = \overline{MR2}/d_2 = 6.71/1.128 = 5.949$ (from a Moving Range chart with CL = 6.71)

(b) Construct a tabular CUSUM for this process, using standardized values h= 8.01 and k = 0.25.

MTB > Stat > Control Charts > Time-Weighted Charts > CUSUM
Enter Subgroup size = 1 and Target = 3200
In CUSUM Options, enter Standard deviation = 5.949, and on the Plan/Type tab, select One-sided type
of CUSUM, and set h = 8.01 and k = 0.25

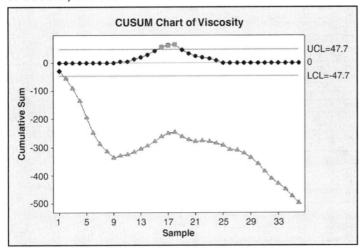

Test Results for CUSUM Chart of Ex9-9Vis
```
TEST. One point beyond control limits.
Test Failed at points:   2, 3, 4, 5, 6, 7, 8, 9, 10, 11, 12, 13, 14, 15, 16, 17,
                        18, 19, 20, 21, 22, 23, 24, 25, 26, 27, 28, 29, 30, 31,
                        32, 33, 34, 35, 36
```

The process signals out of control on the lower side at sample 2 and on the upper side at sample 16.
Assignable causes occurred after startup (sample 2 – 2) and after sample 9 (16 – 7).

(c) Discuss the choice of h and k in part (b) of this problem on CUSUM performance.

Selecting a smaller shift to detect, k = 0.25, should be balanced by a larger control limit, h = 8.01, to give
longer in-control ARLs with shorter out-of-control ARLs.

9.19.

Apply the scale CUSUM discussed in Section 9.1.8 to the data in Exercise 9.1

$$V_i = \left(\sqrt{|y_i|} - 0.822\right)\big/0.349$$

(Note: Solution is in Instructor's Excel data set for Chapter 9)

mu0 =	1050	(from Exercise 9-1)
sigma =	25	(from Exercise 9-1)
delta =	1 sigma	
k =	0.5	
h =	5	

					one-sided upper cusum					one-sided lower cusum			
Obs, i	xi	yi	vi		Si+	N+	OOC?	When?		Si-	N-	OOC?	When?
No FIR					0					0			
1	1045	-0.2	-1.07		0	0				0.574	1		
2	1055	0.2	-1.07		0	0				1.148	2		
3	1037	-0.52	-0.29		0	0				0.937	3		
4	1064	0.56	-0.21		0	0				0.648	4		
5	1095	1.8	1.49		0.989	1				0	0		
6	1008	-1.68	1.36		1.848	2				0	0		
7	1050	0	-2.36		0	0				1.855	1		
8	1087	1.48	1.13		0.631	1				0.225	2		
9	1125	3	2.61		2.738	2				0	0		
10	1146	3.84	3.26		5.498	3	OOC	7		0	0		
11	1139	3.56	3.05		8.049	4	OOC	7		0	0		
12	1169	4.76	3.90		11.445	5	OOC	7		0	0		
13	1151	4.04	3.40		14.349	6	OOC	7		0	0		
14	1128	3.12	2.71		16.555	7	OOC	7		0	0		
15	1238	7.52	5.50		21.557	8	OOC	7		0	0		
16	1125	3	2.61		23.664	9	OOC	7		0	0		
17	1163	4.52	3.74		26.901	10	OOC	7		0	0		
18	1188	5.52	4.38		30.778	11	OOC	7		0	0		
19	1146	3.84	3.26		33.537	12	OOC	7		0	0		
20	1167	4.68	3.84		36.880	13	OOC	7		0	0		

The process is out of control after observation 10 – 3 = 7. Process variability is increasing.

9.21.

Consider a standardized two-sided CUSUM with $k = 0.2$ and $h = 8$. Use Siegmund's procedure to evaluate the in-control ARL performance of this scheme. Find ARL1 for $\delta^* = 0.5$.

In control ARL performance:

$$\delta^* = 0$$

$$\Delta^+ = \delta^* - k = 0 - 0.2 = -0.2$$

$$\Delta^- = -\delta^* - k = -0 - 0.2 = -0.2$$

$$b = h + 1.166 = 8 + 1.166 = 9.166$$

$$ARL_0^+ = ARL_0^- \cong \frac{\exp[-2(-0.2)(9.166)] + 2(-0.2)(9.166) - 1}{2(-0.2)^2} = 430.556$$

$$\frac{1}{ARL_0} = \frac{1}{ARL_0^+} + \frac{1}{ARL_0^-} = \frac{2}{430.556} = 0.005$$

$$ARL_0 = 1/0.005 = 215.23$$

Out of control ARL Performance:

$$\delta^* = 0.5$$

$$\Delta^+ = \delta^* - k = 0.5 - 0.2 = 0.3$$

$$\Delta^- = -\delta^* - k = -0.5 - 0.2 = -0.7$$

$$b = h + 1.166 = 8 + 1.166 = 9.166$$

$$ARL_1^+ = \frac{\exp[-2(0.3)(9.166)] + 2(0.3)(9.166) - 1}{2(0.3)^2} = 25.023$$

$$ARL_1^- = \frac{\exp[-2(-0.7)(9.166)] + 2(-0.7)(9.166) - 1}{2(-0.7)^2} = 381,767$$

$$\frac{1}{ARL_1} = \frac{1}{ARL_1^+} + \frac{1}{ARL_1^-} = \frac{1}{25.023} + \frac{1}{381,767} = 0.040$$

$$ARL_1 = 1/0.040 = 25.02$$

9.23.

Consider the velocity of light data introduced in Exercises 6.69 and 6.70 (Note: earlier printings of the 7th edition state 6.59 and 6.60). Use only the 20 observations in Exercise 6.69 to set up a CUSUM with target value 734.5. Plot all 40 observations from both Exercises 6.69 and 6.70 on this CUSUM. What conclusions can you draw?

▪ **TABLE 6E.27**
Velocity of Light Data for Exercise 6.69

Measurement	Velocity	Measurement	Velocity
1	850	11	850
2	1000	12	810
3	740	13	950
4	980	14	1000
5	900	15	980
6	930	16	1000
7	1070	17	980
8	650	18	960
9	930	19	880
10	760	20	960

▪ **TABLE 6E.28**
Additional Velocity of Light Data for Exercise 6.70

Measurement	Velocity	Measurement	Velocity
21	960	31	800
22	830	32	830
23	940	33	850
24	790	34	800
25	960	35	880
26	810	36	790
27	940	37	900
28	880	38	760
29	880	39	840
30	880	40	800

$\hat{\sigma} = \overline{MR2}/d_2 = 122.6/1.128 = 108.7$ (from the Moving Range chart with CL = 122.6)

$\mu_0 = 734.5; k = 0.5; h = 5; \quad K = k\hat{\sigma} = 0.5(108.7) = 54.35; \quad H = h\hat{\sigma} = 5(108.7) = 543.5$

9.23. continued

MTB > Stat > Control Charts > Time-Weighted Charts > CUSUM
Enter Subgroup size = 1 and Target = 734.5
In CUSUM Options, enter Standard deviation = 108.7, and on the Plan/Type tab, select One-sided type of CUSUM, and set h = 5 and k = 0.5

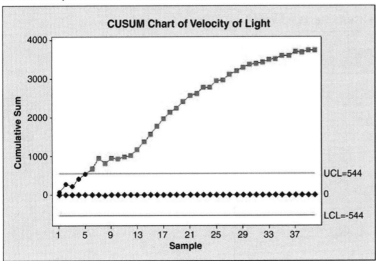

Test Results for CUSUM Chart of Ex6-70Vel
```
TEST. One point beyond control limits.
Test Failed at points:   6, 7, 8, 9, 10, 11, 12, 13, 14, 15, 16, 17, 18, 19, 20,
                         21, 22, 23, 24, 25, 26, 27, 28, 29, 30, 31, 32, 33, 34,
                         35, 36, 37, 38, 39, 40
```

The Individuals I-MR chart, with a centerline at $\bar{x} = 909$, displayed a distinct downward trend in measurements, starting at about sample 18. The CUSUM chart reflects a consistent run above the target value 734.5, from virtually the first sample. There is a distinct signal on both charts, of either a trend/drift or a shit in measurements. The out-of-control signals should lead us to investigate and determine the assignable cause.

9.25.

Rework Exercise 9.1 using an EWMA control chart with $\lambda = 0.1$ and $L = 2.7$. Compare your results to those obtained with the CUSUM.

$\sigma = 25$, CL $= \mu_0 = 1050$, UCL $= 1065.49$, LCL $= 1034.51$

MTB > Stat > Control Charts > Time-Weighted Charts > EWMA

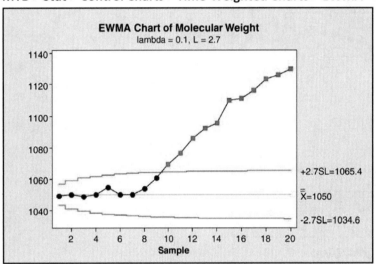

Process exceeds upper control limit at sample 10; the same as the CUSUM chart.

9.27.

Reconsider the data in Exercise 9.4. Set up an EWMA control chart with $\lambda = 0.2$ and $L = 3$ for this process. Interpret the results..

Assume $\sigma = 0.05$. CL $= \mu_0 = 8.02$, UCL $= 8.07$, LCL $= 7.97$

MTB > Stat > Control Charts > Time-Weighted Charts > EWMA

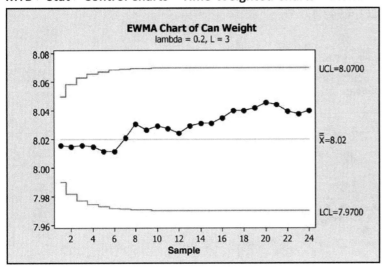

The process is in control.

9.29.

(NOTE: In early printings of the 7th edition, this is listed as Exercise 9.33 in the *Answers to Selected Exercises*)

Reconsider the data in Exercise 9.7. Apply an EWMA control chart to these data using $\lambda = 0.1$ and $L = 2.7$.

$\hat{\sigma} = 12.16$, CL $= \mu_0 = 950$, UCL $= 957.53$, LCL $= 942.47$.

MTB > Stat > Control Charts > Time-Weighted Charts > EWMA

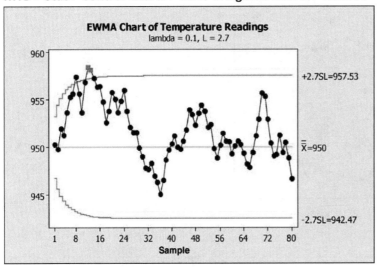

Test Results for EWMA Chart of Ex9.7temp

```
TEST. One point beyond control limits.
Test Failed at points:  12, 13
```

Process is out of control at samples 8 (beyond upper limit, but not flagged on chart), 12 and 13.

9.31.

Reconsider the data in Exercise 9.8. Set up and apply an EWMA control chart to these data using $\lambda = 0.05$ and $L = 2.6$.

$\hat{\sigma} = 5.634$, CL = μ_0 = 175, UCL = 177.30, LCL = 172.70.

MTB > Stat > Control Charts > Time-Weighted Charts > EWMA

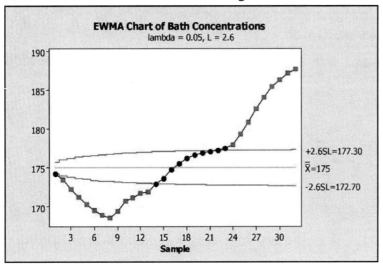

Process is out of control. The process average of $\hat{\mu} = 183.594$ is too far from the process target of μ_0 = 175 for the process variability. The data is grouped into three increasing levels.

9.35.

Analyze the data in Exercise 9.4 using a moving average control chart with $w = 5$. Compare the results obtained with the cumulative sum control chart in Exercise 9.4.

$w = 5$, $\mu_0 = 8.02$, $\sigma = 0.05$, CL = 8.02, UCL = 8.087, LCL = 7.953

MTB > Stat > Control Charts > Time-Weighted Charts > Moving Average
Subgroup size = 1, Length of MA = 5
In MA Options, mean = 8.02, standard deviation = 0.05

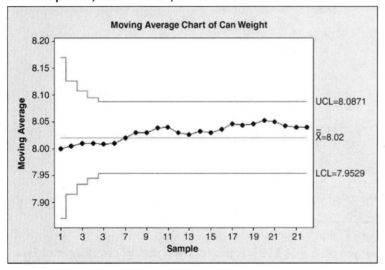

The process is in control, the same result as for Exercise 9.4.

9.43.

An EWMA control chart uses $\lambda = 0.4$. How wide will the limits be on the Shewhart control chart, expressed as a multiple of the width of the steady-state EWMA limits?

For EWMA, steady-state limits are $\pm L\sigma\sqrt{\lambda/(2-\lambda)}$

For Shewhart, steady-state limits are $\pm k\sigma$

$$k\sigma = L\sigma\sqrt{\lambda/(2-\lambda)}$$
$$k = L\sqrt{0.4/(2-0.4)}$$
$$k = 0.5L$$

Chapter 10

Other Univariate Statistical Process-Monitoring and Control Techniques

LEARNING OBJECTIVES

After completing this chapter you should be able to:

1. Set up and use $\bar{x}$ and R control charts for short production runs
2. Know how to calculate modified limits for the Shewhart $\bar{x}$ control chart
3. Know how to set up and use an acceptance control chart
4. Use group control charts for multiple-stream processes, and understand the alternative procedures that are available
5. Understand the sources and effects of autocorrelation on standard control charts
6. Know how to use model-based residuals control charts for autocorrelated data
7. Know how to use the batch means control chart for autocorrelated data
8. Know what the Cuscore control chart can be used for
9. Know how change point methods relate to statistical process monitoring techniques
10. Understand the practical reason behind the use of adaptive control charts
11. Understand the basis of economic principles of control chart design
12. Know how control charts can be used for monitoring processes whose output is a profile

IMPORTANT TERMS AND CONCEPTS

Acceptance control charts	First-order moving average model
Adaptive (SPC) control charts	Group control charts (GCC)
Autocorrelated process data	Health care applications of control charts
Autocorrelation function	Impact of autocorrelation on control charts
Autoregressive integrated moving average (ARIMA) models	Modified control charts
Average run length	Multiple-stream processes
Bernoulli processes	Positive autocorrelation
Changepoint model for process monitoring	Profile monitoring
Control charts on residuals	Sample autocorrelation function

Cuscore statistics and control charts

Deviation from nominal (DNOM) control charts

First-order autoregressive model

First-order integrated moving average model

First-order mixed model

Second-order autoregressive model

Shewhart process model

Standardized $\bar{x}$ and R control charts

Time series model

Unweighted batch means (UBM) control charts

EXERCISES

Note: Many of the exercises in this chapter were solved with Microsoft® Excel® 2010.

10.1.

Use the data in Table 10E.1 to set up short-run $\bar{x}$ and R charts using the DNOM approach. The nominal dimensions for each part are $T_A = 100$, $T_B = 60$, $T_C = 75$, and $T_D = 50$.

■ **TABLE 10E.1**

Data for Exercise 10.1

Sample Number	Part Type	M_1	M_2	M_3
1	A	105	102	103
2	A	101	98	100
3	A	103	100	99
4	A	101	104	97
5	A	106	102	100
6	B	57	60	59
7	B	61	64	63
8	B	60	58	62
9	C	73	75	77
10	C	78	75	76
11	C	77	75	74
12	C	75	72	79
13	C	74	75	77
14	C	73	76	75
15	D	50	51	49
16	D	46	50	50
17	D	51	46	50
18	D	49	50	53
19	D	50	52	51
20	D	53	51	50

$\hat{\sigma}_A = 2.530, n_A = 15, \hat{\mu}_A = 101.40$; $\hat{\sigma}_B = 2.297, n_B = 9, \hat{\mu}_B = 60.444$;

$\hat{\sigma}_C = 1.815, n_C = 18, \hat{\mu}_C = 75.333$; $\hat{\sigma}_D = 1.875, n_D = 18, \hat{\mu}_D = 50.111$

10.1. continued

Standard deviations are approximately the same, so the DNOM chart can be used.

$\bar{R} = 3.8, \hat{\sigma} = 2.245, n = 3$

$\bar{x}$ chart: CL = 0.55, UCL = 4.44, LCL = −3.34

R chart: CL = 3.8, UCL = $D_4\bar{R}$ = 2.574 (3.8) = 9.78, LCL = 0

Stat > Control Charts > Variables Charts for Subgroups > Xbar-R Chart

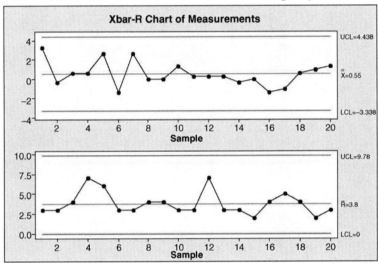

Process is in control, with no samples beyond the control limits or unusual plot patterns.

10.5.

A machine has four heads. Samples of $n= 3$ units are selected from each head, and the $\bar{x}$ and R values for an important quality characteristic are computed. The data are shown in Table10E.4. Set up group control charts for this process.

■ TABLE 10E.4
Data for Exercise 10.5

Sample Number	Head 1 $\bar{x}$	1 R	Head 2 $\bar{x}$	2 R	Head 3 $\bar{x}$	3 R	Head 4 $\bar{x}$	4 R
1	53	2	54	1	56	2	55	3
2	51	1	55	2	54	4	54	4
3	54	2	52	5	53	3	57	2
4	55	3	54	3	52	1	51	5
5	54	1	50	2	51	2	53	1
6	53	2	51	1	54	2	52	2
7	51	1	53	2	58	5	54	1
8	52	2	54	4	51	2	55	2
9	50	2	52	3	52	1	51	3
10	51	1	55	1	53	3	53	5
11	52	3	57	2	52	4	55	1
12	51	2	55	1	54	2	58	2
13	54	4	58	2	51	1	53	1
14	53	1	54	4	50	3	54	2
15	55	2	52	3	54	2	52	6
16	54	4	51	1	53	2	58	5
17	53	3	50	2	57	1	53	1
18	52	1	49	1	52	1	49	2
19	51	2	53	3	51	2	50	3
20	52	4	52	2	50	3	52	2

(Note: Solution is in Student's Excel data set for Chapter 10)

Grand Avg =	52.988
Avg R =	2.338
s =	4 heads
n =	3 units
A2 =	1.023
D3 =	0
D4 =	2.574
Xbar UCL =	55.379
Xbar LCL =	50.596
R UCL =	6.017
R LCL =	0.000

10.5. continued

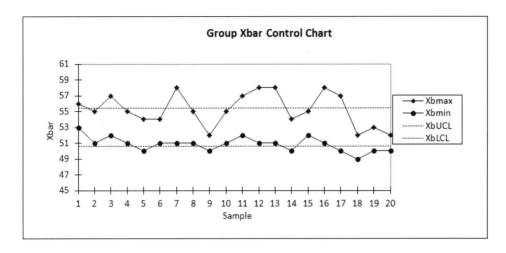

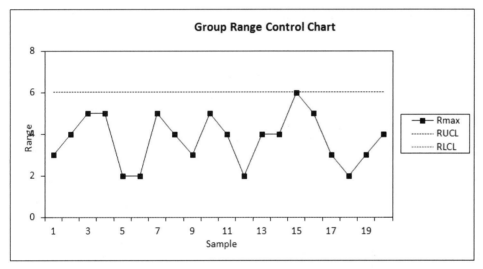

There is no situation where one single head gives the maximum or minimum value of $\bar{x}$ six times in a row. There are many values of $\bar{x}$ max and $\bar{x}$ min that are outside the control limits, so the process is out-of-control. The assignable cause affects all heads, not just a specific one.

10.7.

Reconsider the data in Exercises 10.5 and 10.6. Suppose the process measurements are individual data values, not subgroup averages.

 (a) Use observations 1-20 in Exercise 10.5 to construct appropriate group control charts.

 (Note: Solution is in Student's Excel data set for Chapter 10)

Grand Avg =	52.988	
Avg MR =	2.158	
s =	4	heads
n =	2	units
d2 =	1.128	
D3 =	0	
D4 =	3.267	
Xbar UCL =	58.727	
Xbar LCL =	47.248	
R UCL =	7.050	
R LCL =	0.000	

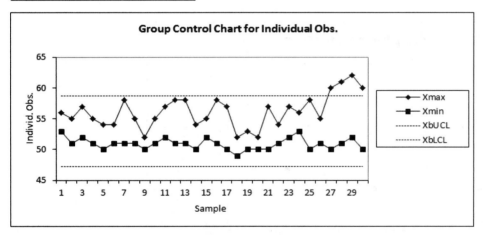

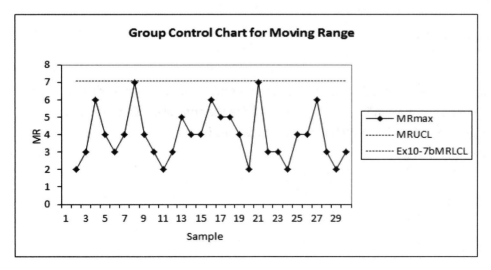

See the discussion in Exercise 10.5.

10.7. continued

(b) Plot observations 21-30 from Exercise 10.6 the charts from part (a). Discuss your findings.

(Note: Solution is in Student's Excel data set for Chapter 10)

Grand Avg =	52.988	
Avg MR =	2.158	
s =	4	heads
n =	2	units
d2 =	1.128	
D3 =	0	
D4 =	3.267	
Xbar UCL =	58.727	
Xbar LCL =	47.248	
R UCL =	7.050	
R LCL =	0.000	

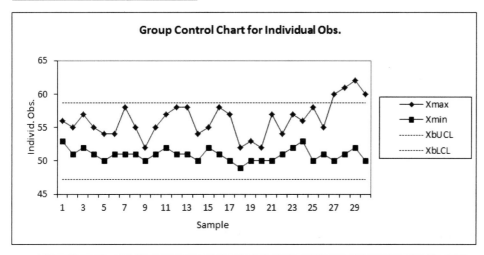

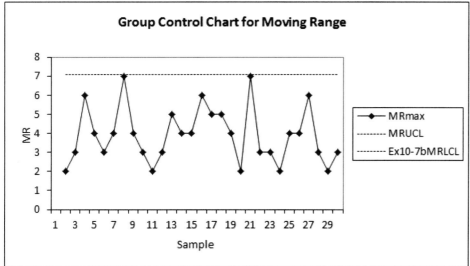

The last four samples from Head 4 remain the maximum of all heads; indicating a potential process change.

10.7. continued

(c) Using observations 1-20, construct an individuals control chart using the average of the readings on all four heads as an individual measurement and an s control chart using the individual measurements on each head. Discuss how these charts function relative to the group control chart.

Stat > Control Charts > Variables Charts for Subgroups > Xbar-S Chart
Note: Use "Sbar" as the method for estimating standard deviation.

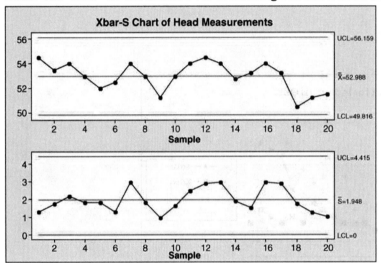

Failure to recognize the multiple stream nature of the process had led to control charts that fail to identify the out-of-control conditions in this process.

10.7. continued

(d) Plot observations 21-30 on the control charts from part (c). Discuss your findings.

Stat > Control Charts > Variables Charts for Subgroups > Xbar-S Chart
Note: Use "Sbar" as the method for estimating standard deviation.

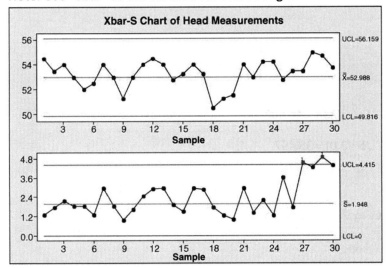

Test Results for S Chart of Ex10.7X1, ..., Ex10.7X4

```
TEST 1. One point more than 3.00 standard deviations from center line.
Test Failed at points:   27, 29
```

Only the S chart gives any indication of out-of-control process.

10.9.

A sample of five units is taken from a process every half hour. It is known that the process standard deviation is in control with $\sigma = 2.0$. The $\bar{x}$ values for the last 20 samples are shown in Table 10E.7. Specifications on the product are 40 ± 8.

■ TABLE 10E.7
Data for Exercise 10.9

Sample Number	$\bar{x}$	Sample Number	$\bar{x}$
1	41.5	11	40.6
2	42.7	12	39.4
3	40.5	13	38.6
4	39.8	14	42.5
5	41.6	15	41.8
6	44.7	16	40.7
7	39.6	17	42.8
8	40.2	18	43.4
9	41.4	19	42.0
10	43.9	20	41.9

10.9. continued

(a) Set up a modified control chart on this process. Use three-sigma limits on the chart and assume that the largest fraction nonconforming that is tolerable is 0.1%

3-sigma limits:

$n = 5, \delta = 0.001, Z_\delta = Z_{0.001} = 3.090$

$\text{USL} = 40 + 8 = 48, \text{LSL} = 40 - 8 = 32$

$\text{UCL} = \text{USL} - \left(Z_\delta - 3/\sqrt{n}\right)\sigma = 48 - \left(3.090 - 3/\sqrt{5}\right)(2.0) = 44.503$

$\text{LCL} = \text{LSL} + \left(Z_\delta - 3/\sqrt{n}\right)\sigma = 32 + \left(3.090 - 3/\sqrt{5}\right)(2.0) = 35.497$

Graph > Time Series Plot > Simple
Note: Reference lines have been used set to the control limit values.

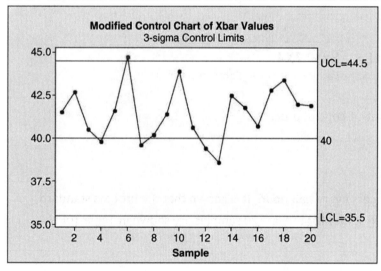

Process is out of control at sample #6.

10.9. continued

(b) Reconstruct the chart in part (a) using two-sigma limits. Is there any difference in the analysis of the data?

2-sigma limits:

$$UCL = USL - \left(Z_\delta - 2/\sqrt{n}\right)\sigma = 48 - \left(3.090 - 2/\sqrt{5}\right)(2.0) = 43.609$$

$$LCL = LSL + \left(Z_\delta - 2/\sqrt{n}\right)\sigma = 32 + \left(3.090 - 2/\sqrt{5}\right)(2.0) = 36.391$$

Graph > Time Series Plot > Simple
Note: Reference lines have been used set to the control limit values.

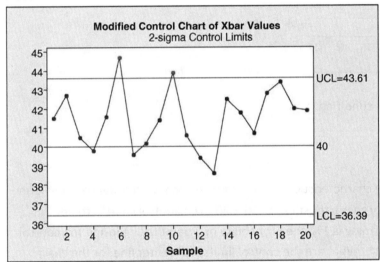

With 3-sigma limits, sample #6 exceeds the UCL, while with 2-sigma limits both samples #6 and #10 exceed the UCL.

(c) Suppose that if the true process fraction nonconforming is 5%, we would like to detect this condition with probability 0.95. Construct the corresponding acceptance control chart.

$$\gamma = 0.05, Z_\gamma = Z_{0.05} = 1.645$$

$$1 - \beta = 0.95, Z_\beta = Z_{0.05} = 1.645$$

$$UCL = USL - \left(Z_\gamma + z_\beta/\sqrt{n}\right)\sigma = 48 - \left(1.645 + 1.645/\sqrt{5}\right)(2.0) = 43.239$$

$$LCL = LSL + \left(Z_\gamma + z_\beta/\sqrt{n}\right)\sigma = 32 + \left(1.645 + 1.645/\sqrt{5}\right)(2.0) = 36.761$$

10.9.(c) continued

Graph > Time Series Plot > Simple

Note: Reference lines have been used set to the control limit values.

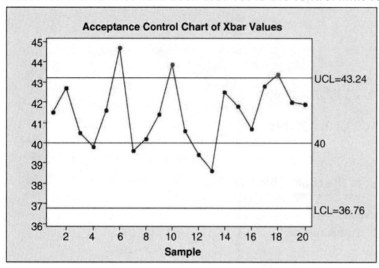

Sample #18 also signals an out-of-control condition.

10.13.

An $\bar{x}$ chart is to be designed for a quality characteristic assumed to be normal with a standard deviation of 4. Specifications on the product quality characteristics are 50 ± 20. The control chart is to be designed so that if the fraction nonconforming is 1%, the probability of a point falling inside the control limits will be 0.99. The sample size is $n = 4$. What are the control limits and center line for the chart?

$\delta = 0.01$, $Z_\delta = 2.326$; $1 - \alpha = 0.995$, $\alpha = 0.005$, $Z_\alpha = 2.576$

$$\text{UCL} = \text{USL} - \left(Z_\delta - Z_\alpha / \sqrt{n}\right)\sigma = (50 + 20) - \left(2.326 - 2.576 / \sqrt{4}\right)(4) = 65.848$$

$$\text{LCL} = \text{LSL} + \left(Z_\delta - Z_\alpha / \sqrt{n}\right)\sigma = (50 - 20) + \left(2.326 - 2.576 / \sqrt{4}\right)(4) = 34.152$$

10.15.

A normally distributed quality characteristic is controlled by $\bar{x}$ and R charts having the following parameters ($n = 4$, both charts are in control):

R Chart	$\bar{x}$ Chart
UCL = 18.795	UCL = 626.00
Center line = 8.236	Center line = 620.00
LCL = 0	LCL = 614.00

(a) What is the estimated standard deviation of the quality characteristic x?

$$\hat{\sigma}_x = \bar{R}/d_2 = 8.236/2.059 = 4.000$$

(b) If specifications are 610 ± 15, what is your estimate of the fraction of nonconforming material produced by this process when it is in control at the given level?

$$\hat{p} = Pr\{x < LSL\} + Pr\{x > USL\}$$
$$= Pr\{x < 595\} + \left[1 - Pr\{x \leq 625\}\right]$$
$$= \Phi\left(\frac{595 - 620}{4.000}\right) + \left[1 - \Phi\left(\frac{625 - 620}{4.000}\right)\right]$$
$$= 0.0000 + \left[1 - 0.8944\right]$$
$$= 0.1056$$

(c) Suppose you wish to establish a modified $\bar{x}$ chart to substitute for the original $\bar{x}$ chart. The process mean is to be controlled so that the fraction nonconforming is less than 0.005. The probability of type I error is to be 0.01. What control limits do you recommend?

$$\delta = 0.005, Z_\delta = Z_{0.005} = 2.576$$
$$\alpha = 0.01, Z_\alpha = Z_{0.01} = 2.326$$
$$UCL = USL - \left(Z_\delta - Z_\alpha/\sqrt{n}\right)\sigma = 625 - \left(2.576 - 2.326/\sqrt{4}\right)4 = 619.35$$
$$LCL = LSL + \left(Z_\delta - Z_\alpha/\sqrt{n}\right)\sigma = 595 + \left(2.576 - 2.326/\sqrt{4}\right)4 = 600.65$$

10.17.

Consider the molecular weight data in Exercise 10.16. Construct a CUSUM control chart on the residuals from the model you fit to the data in part (c) of that exercise.

Let $\mu_0 = 0$, $\delta = 1$ sigma, $k = 0.5$, $h = 5$.

Stat > Control Charts > Time-Weighted Charts > CUSUM

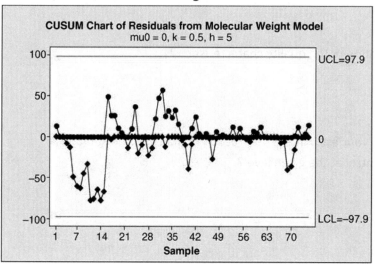

No observations exceed the control limit. The process is in control.

10.19.

Set up a moving center-line EWMA control chart for the molecular weight data in Exercise 10.16. Compare it to the residual control chart in Exercise 10.16, part (c).

To find the optimal λ, fit an ARIMA (0,1,1) (= EWMA = IMA(1,1)):

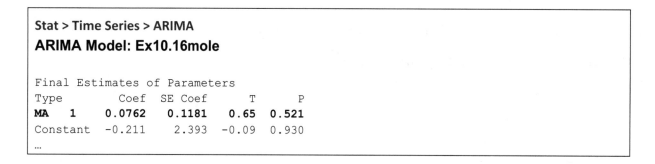

```
Stat > Time Series > ARIMA
ARIMA Model: Ex10.16mole


Final Estimates of Parameters
Type         Coef   SE Coef      T      P
MA    1    0.0762    0.1181   0.65  0.521
Constant  -0.211     2.393   -0.09  0.930
...
```

10.19. continued

$\lambda = 1 - MA1 = 1 - 0.0762 = 0.9238$ and $\hat{\sigma} = \overline{MR}/d_2 = 17.97/1.128 = 15.93$

(Note: Solution is in Student's Excel data set for Chapter 10)

t	xt	zt	CL	UCL	LCL	OOC?
0		2000.947				
1	2048	2044.415	2000.947	2048.749	1953.145	No
2	2025	2026.479	2044.415	2092.217	1996.613	No
3	2017	2017.722	2026.479	2074.281	1978.677	No
4	1995	1996.731	2017.722	2065.524	1969.920	No
5	1983	1984.046	1996.731	2044.533	1948.929	No
6	1943	1946.128	1984.046	2031.848	1936.244	No
7	1940	1940.467	1946.128	1993.930	1898.326	No
8	1947	1946.502	1940.467	1988.269	1892.665	No
9	1972	1970.057	1946.502	1994.304	1898.700	No
10	1983	1982.014	1970.057	2017.859	1922.255	No
11	1935	1938.582	1982.014	2029.816	1934.212	No
12	1948	1947.282	1938.582	1986.384	1890.780	No
13	1966	1964.574	1947.282	1995.084	1899.480	No
14	1954	1954.806	1964.574	2012.376	1916.772	No
15	1970	1968.842	1954.806	2002.608	1907.004	No
16	2039	2033.654	1968.842	2016.644	1921.040	above UCL
17	2015	2016.421	2033.654	2081.456	1985.852	No

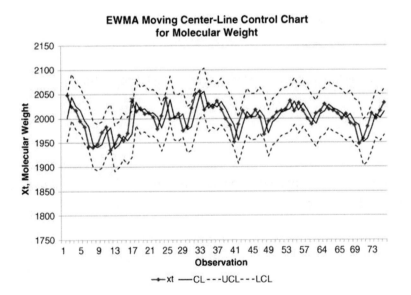

Observation 16 exceeds the upper control limit compared to one out-of-control signal at observation 16 on the Individuals control chart.

10.21.

Consider the concentration data in Exercise 10.20. Construct a CUSUM chart on the residuals from the model you fit in part (c) of that exercise.

Let $\mu_0 = 0$, $\delta = 1$ sigma, $k = 0.5$, $h = 5$.

Stat > Control Charts > Time-Weighted Charts > CUSUM

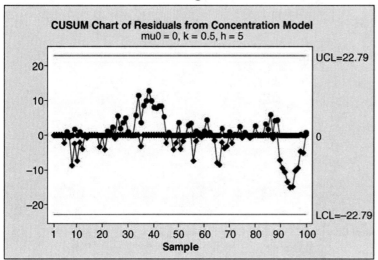

No observations exceed the control limit. The residuals are in control, and the AR(1) model for concentration should be a good fit.

10.23.

Set up a moving center line EWMA control chart for the concentration data in Exercise 10.20. Compare it to the residuals control chart in Exercise 10.20, part (c).

10.23. continued

To find the optimal λ, fit an ARIMA (0,1,1) (= EWMA = IMA(1,1)):

Stat > Time Series > ARIMA

ARIMA Model: Ex10.20conc

...

```
Final Estimates of Parameters
Type          Coef  SE Coef      T      P
MA    1     0.2945   0.0975   3.02  0.003
Constant  -0.0452   0.3034  -0.15  0.882
```

...

$\lambda = 1 - MA1 = 1 - 0.2945 = 0.7055$ and $\hat{\sigma} = \overline{MR}/d_2 = 3.64/1.128 = 3.227$

(Note: Solution is in Student's Excel data set for Chapter 10)

lamda =	0.706	sigma^ =	3.23			
t	xt	zt	CL	UCL	LCL	OOC?
0		200.010				
1	204	202.825	200.010	209.691	190.329	0
2	202	202.243	202.825	212.506	193.144	0
3	201	201.366	202.243	211.924	192.562	0
4	202	201.813	201.366	211.047	191.685	0
5	197	198.418	201.813	211.494	192.132	0
6	201	200.239	198.418	208.099	188.737	0
7	198	198.660	200.239	209.920	190.558	0
8	188	191.139	198.660	208.341	188.979	below LCL
9	195	193.863	191.139	200.820	181.458	0

...

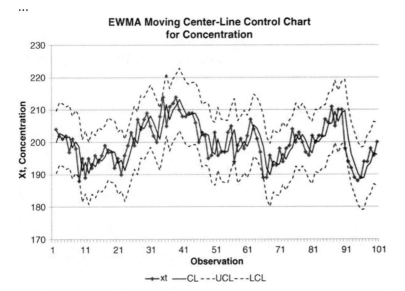

EWMA Moving Center-Line Control Chart for Concentration

The control chart of concentration data signals out of control at three observations (8, 56, 90).

10.25.

Consider the temperature data in exercise 10.24. Set up a CUSUM control chart on the residuals from the model you fit to the data in part (c) of that exercise. Compare it to the individuals chart you constructed using the residuals.

MTB > Stat > Control Charts > Time-Weighted Charts > CUSUM

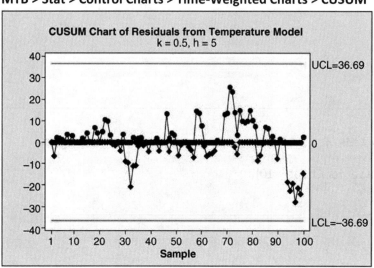

No observations exceed the control limits. The residuals are in control, indicating the process is in control. This is the same conclusion as applying an Individuals control chart to the residuals.

10.27.

Set up a moving center line EWMA control chart for the temperature data in Exercise 10.24. Compare it to the residuals control chart in Exercise 10.24, part (c).

To find the optimal λ, fit an ARIMA (0,1,1) (= EWMA = IMA(1,1)):

Stat > Time Series > ARIMA
ARIMA Model: Ex10.24temp

```
...
Final Estimates of Parameters
Type          Coef   SE Coef      T       P
MA    1     0.0794   0.1019    0.78   0.438
Constant   -0.0711   0.6784   -0.10   0.917
...
```

10.27. continued

$\lambda = 1 - MA1 = 1 - 0.0794 = 0.9206$

$\hat{\sigma} = \overline{MR}/d_2 = 5.75/1.128 = 5.0975$ (from a Moving Range chart with CL = 5.75)

(Note: Solution is in Student's Excel data set for Chapter 10)

		lambda =	0.921	sigma^ =	5.098	
t	xt	zt	CL	UCL	LCL	OOC?
0		506.520				
1	491	492.232	506.520	521.813	491.227	below LCL
2	482	482.812	492.232	507.525	476.940	0
3	490	489.429	482.812	498.105	467.520	0
4	495	494.558	489.429	504.722	474.137	0
5	499	498.647	494.558	509.850	479.265	0
6	499	498.972	498.647	513.940	483.355	0
7	507	506.363	498.972	514.265	483.679	0
8	503	503.267	506.363	521.655	491.070	0
9	510	509.465	503.267	518.560	487.974	0
10	509	509.037	509.465	524.758	494.173	0
11	510	509.924	509.037	524.330	493.744	0

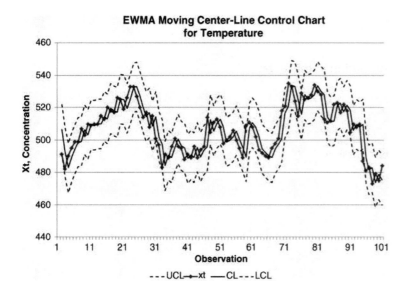

A few observations exceed the upper limit (46, 58, 69) and the lower limit (1, 94), similar to the two out-of-control signals on the Individuals control chart (71, 94).

10.29.

The viscosity of a chemical product is read every two minutes. Some data from this process are shown in Table 10E.11 (read down, the across from left to right).

■ TABLE 10E.11
Chemical Product Viscosity

29.330	33.220	27.990	24.280
19.980	30.150	24.130	22.690
25.760	27.080	29.200	26.600
29.000	33.660	34.300	28.860
31.030	36.580	26.410	28.270
32.680	29.040	28.780	28.170
33.560	28.080	21.280	28.580
27.500	30.280	21.710	30.760
26.750	29.350	21.470	30.620
30.550	33.600	24.710	20.840
28.940	30.290	33.610	16.560
28.500	20.110	36.540	25.230
28.190	17.510	35.700	31.790
26.130	23.710	33.680	32.520
27.790	24.220	29.290	30.280
27.630	32.430	25.120	26.140
29.890	32.440	27.230	19.030
28.180	29.390	30.610	24.340
26.650	23.450	29.060	31.530
30.010	23.620	28.480	31.950
30.800	28.120	32.010	31.680
30.450	29.940	31.890	29.100
36.610	30.560	31.720	23.150
31.400	32.300	29.090	26.740
30.830	31.580	31.920	32.440

(a) Is there a serious problem with autocorrelation in these data?

Stat > Time Series > Autocorrelation

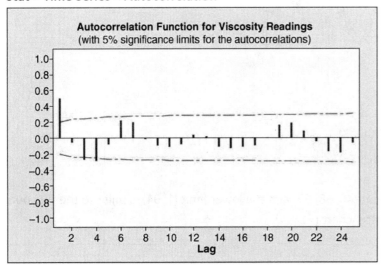

10.29.(a) continued

```
Autocorrelation Function: Ex10.29Vis
Lag        ACF        T      LBQ
  1    0.494137    4.94    25.16
  2   -0.049610   -0.41    25.41
  3   -0.264612   -2.17    32.78
  4   -0.283150   -2.22    41.29
  5   -0.071963   -0.54    41.85
...
```

$r_1 = 0.49$, indicating a strong positive correlation at lag 1. There is a serious problem with autocorrelation in viscosity readings.

 (b) Set up a control chart for individuals with a moving range used to estimate process variability. What conclusion can you draw from this chart?

Stat > Control Charts > Variables Charts for Individuals > Individuals

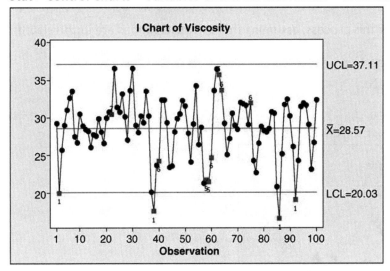

10.29.(b) continued

Test Results for I Chart of Ex10.29Vis

TEST 1. One point more than 3.00 standard deviations from center line.
Test Failed at points: 2, 38, 86, 92
TEST 5. 2 out of 3 points more than 2 standard deviations from center line (on
    one side of CL).
Test Failed at points: 38, 58, 59, 63, 86
TEST 6. 4 out of 5 points more than 1 standard deviation from center line (on
    one side of CL).
Test Failed at points: 40, 60, 64, 75
TEST 7. 15 points within 1 standard deviation of center line (above and below
    CL).
Test Failed at points: 22
TEST 8. 8 points in a row more than 1 standard deviation from center line
    (above and below CL).
Test Failed at points: 64

Process is out of control, violating many of the tests for special causes. The viscosity measurements appear to wander over time.

(c) Design a CUSUM control scheme for this process, assuming that the observations are uncorrelated. How does the CUSUM perform?

Let target = μ_0 = 28.569

MTB > Stat > Control Charts > Time-Weighted Charts > CUSUM

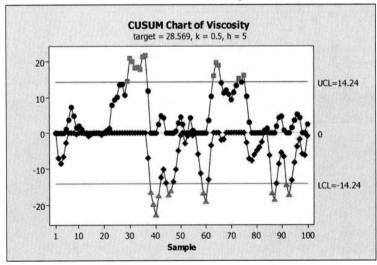

Several observations are out of control on both the lower and upper sides.

10.29. continued

(d) Set up an EWMA control chart with $\lambda = 0.15$ for the process. How does this chart perform?

MTB > Stat > Control Charts > Time-Weighted Charts > EWMA

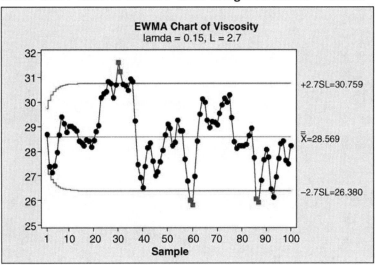

The process is not in control. There are wide swings in the plot points and several are beyond the control limits.

(e) Set up a moving center line EWMA scheme for these data.

To find the optimal λ, fit an ARIMA (0,1,1) (= EWMA = IMA(1,1)):

```
Stat > Time Series > ARIMA
ARIMA Model: Ex10.29Vis
...
Final Estimates of Parameters
Type          Coef   SE Coef      T      P
MA    1    -0.1579   0.1007   -1.57  0.120
Constant   0.0231   0.4839    0.05  0.962
```

$\lambda = 1 - MA1 = 1 - (-0.1579) = 1.1579$

$\hat{\sigma} = \overline{MR}/d_2 = 3.21/1.128 = 2.8457$ (from a Moving Range chart with CL = 5.75)

10.29.(e) continued

(Note: Solution is in Student's Excel data set for Chapter 10)

I	Xi	Zi	CL	UCL	LCL	OOC?
0		28.479				
1	29.330	29.464	28.479	37.022	19.937	0
2	19.980	18.482	29.464	38.007	20.922	below LCL
3	25.760	26.909	18.482	27.025	9.940	0
4	29.000	29.330	26.909	35.452	18.367	0
5	31.030	31.298	29.330	37.873	20.788	0
6	32.680	32.898	31.298	39.841	22.756	0
7	33.560	33.665	32.898	41.441	24.356	0
8	27.500	26.527	33.665	42.207	25.122	0
9	26.750	26.785	26.527	35.069	17.984	0
10	30.550	31.144	26.785	35.328	18.243	0
11	28.940	28.592	31.144	39.687	22.602	0

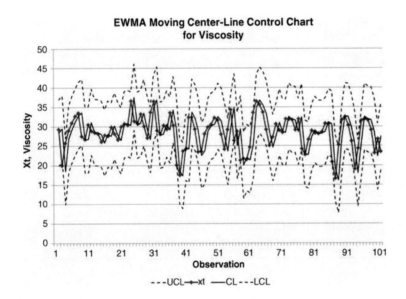

EWMA Moving Center-Line Control Chart for Viscosity

A few observations exceed the upper limit (87) and the lower limit (2, 37, 55, 85).

(f) Suppose that a reasonable model for the viscosity data is an AR(2) model. How could this model be used to assist in the development of a statistical process-control procedure for viscosity? Set up an appropriate control chart, and use it to assess the current state of statistical control.

10.29.(f) continued

Stat > Time Series > ARIMA
ARIMA Model: Ex10.29Vis

...
```
Final Estimates of Parameters
Type          Coef  SE Coef      T       P
AR   1      0.7193   0.0923    7.79   0.000
AR   2     -0.4349   0.0922   -4.72   0.000
Constant   20.5017   0.3278   62.54   0.000
Mean       28.6514   0.4581
...
```

Stat > Control Charts > Variables Charts for Individuals > Individuals

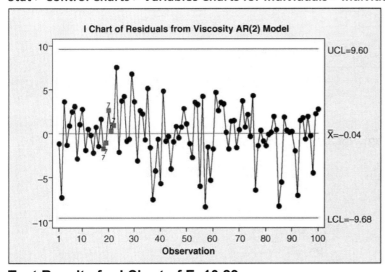

Test Results for I Chart of Ex10.29res
```
TEST 7. 15 points within 1 standard deviation of center line (above and below
     CL).
Test Failed at points:  18, 19, 20, 21, 22
```

The model residuals signal a potential issue with viscosity around observation 20. Otherwise the process appears to be in control, with a good distribution of points between the control limits and no patterns.

10.31.

An $\bar{x}$ chart is used to maintain current control of a process. The cost parameters are a_1 = $0.50, a_2 = $0.10, a_3 = $25, a'_3 = $50, and a_4 = $100. A single assignable cause of magnitude δ = 2 occurs, and the duration of the process in control is an exponential random variable with mean 100 h. Sampling and testing require 0.05 h, and it takes 2 h to locate the assignable cause. Assume that equation 10-31 is the appropriate process model.

λ = 0.01/hr or $1/\lambda$ = 100hr; δ = 2.0

a_1 = $0.50/sample; a_2 = $0.10/unit; a'_3 = $50; a_3 = $25; a_4 = $100/hr

g = 0.05hr/sample; D = 2hr

(a) Evaluate the cost of the arbitrary control chart design n = 5, k = 3, and h= 1.

(Note: Solution is in Student's Excel data set for Chapter 10)

Let α = 0.0027

$$\beta = \Phi\left(\frac{(\mu_0 + k\sigma/\sqrt{n}) - (\mu_0 + \delta\sigma)}{\sigma/\sqrt{n}}\right) - \Phi\left(\frac{(\mu_0 - k\sigma/\sqrt{n}) - (\mu_0 + \delta\sigma)}{\sigma/\sqrt{n}}\right)$$

$$= \Phi\left(k - \delta\sqrt{n}\right) - \Phi\left(-k - \delta\sqrt{n}\right)$$

$$= \Phi\left(3 - 2\sqrt{5}\right) - \Phi\left(-3 - 2\sqrt{5}\right) = \Phi(-1.472) - \Phi(-7.472) = 0.0705 - 0.0000 = 0.0705$$

$$\tau \cong \frac{h}{2} - \frac{\lambda h^2}{12} = \frac{1}{2} - \frac{0.01(1^2)}{12} = 0.4992$$

$$\frac{\alpha e^{-\lambda h}}{\left(1 - e^{-\lambda h}\right)} \cong \frac{\alpha}{\lambda h} = \frac{0.0027}{0.01(1)} = 0.27$$

$E(L)$ = $4.12/hr

(b) Evaluate the cost of the arbitrary control chart design n = 5, k = 3, and h= 0.5.

(Note: Solution is in Student's Excel data set for Chapter 10)

Let α = 0.0027, β = 0.0705

$$\tau \cong \frac{h}{2} - \frac{\lambda h^2}{12} = \frac{0.5}{2} - \frac{0.01(0.5^2)}{12} = 0.2498$$

$$\frac{\alpha e^{-\lambda h}}{\left(1 - e^{-\lambda h}\right)} \cong \frac{\alpha}{\lambda h} = \frac{0.0027}{0.01(0.5)} = 0.54$$

$E(L)$ = $4.98/hr

10.31. continued

(c) Determine the economically optimum design.

$n = 5$, $k_{opt} = 3.080$, $h_{opt} = 1.368$, $\alpha = 0.00207$, $1 - \beta = 0.918$

$E(L) = \$4.01392/hr$

10.33.

An $\bar{x}$ chart is used to maintain current control of a process. The cost parameters are $a_1 = \$2$, $a_2 = \$0.50$, $a_3 = \$50$, $a\boxempty_3 = \$75$, and $a_4 = \$200$. A single assignable cause of magnitude $\delta = 1$ occurs, and the duration of the process in control is an exponential random variable with mean 100 h. Sampling and testing require 0.05 h, and it takes 1 h to locate the assignable cause.

$\lambda = 0.01/hr$ or $1/\lambda = 100hr$, $\delta = 2.0$
$a_1 = \$2/sample$; $a_2 = \$0.50/unit$; $a'_3 = \$75$; $a_3 = \$50$; $a_4 = \$200/hr$
$g = 0.05$ hr/sample, $D = 1$ hr

(a) Evaluate the cost of the arbitrary $\bar{x}$ chart design $n = 5$, $k = 3$, and $h = 0.05$.

(Note: Solution is in Student's Excel data set for Chapter 10)

Assume that equation 10-31 is the appropriate process model.

Let $\alpha = 0.0027$

$$\beta = \Phi\left(k - \delta\sqrt{n}\right) - \Phi\left(-k - \delta\sqrt{n}\right)$$

$$= \Phi\left(3 - 1\sqrt{5}\right) - \Phi\left(-3 - 1\sqrt{5}\right) = \Phi(-1.472) - \Phi(-7.472) = 0.775 - 0.0000 = 0.775$$

$$\tau \cong \frac{h}{2} - \frac{\lambda h^2}{12} = \frac{0.5}{2} - \frac{0.01(0.5^2)}{12} = 0.2498$$

$$\frac{\alpha e^{-\lambda h}}{\left(1 - e^{-\lambda h}\right)} \cong \frac{\alpha}{\lambda h} = \frac{0.0027}{0.01(0.5)} = 0.54$$

$E(L) = \$16.17/hr$

(b) Find the economically optimum design.

$n = 10$, $k_{opt} = 2.240$, $h_{opt} = 2.489018$, $\alpha = 0.025091$, $1 - \beta = 0.8218083$

$E(L) = \$10.39762/hr$

Chapter 11

Multivariate Process Monitoring and Control

Learning Objectives

After completing this chapter you should be able to:
1. Understand why applying several univariate control carts simultaneously to a set of related quality characteristics may be an unsatisfactory monitoring procedure
2. Understand how the multivariate normal distribution is used as a model for multivariate process data
3. Know how to estimate the mean vector and covariance matrix from a sample of multivariate observations
4. Know how to set up and use a chi-square control chart
5. Know how to set up and use the Hotelling T^2 control chart
6. Know how to set up and use the multivariate exponentially weighted moving average (MEWMA) control chart
7. Know how to use multivariate control charts for individual observations
8. Know how to find the phase I and phase II limits for multivariate control charts
9. Use control charts for monitoring multivariate variability
10. Understand the basis of the regression adjustment procedure and know how to apply regression adjustment in process monitoring
11. Understand the basis of principal components and know how to apply principal component sin process monitoring

IMPORTANT TERMS AND CONCEPTS

Average run length (ARL)
Cascade process
Chi-square control chart
Control ellipse
Covariance matrix
Hotelling T^2 control chart
Hotelling T^2 individuals control chart
Hotelling T^2 subgrouped data control chart
Latent structure models
Matrix of scatter plots
Monitoring multivariate variability
Multivariate EWMA control chart
Multivariate normal distribution

Multivariate quality control process monitoring
Partial least squares (PLS)
Phase I control limits
Phase II control limits
Principal component scores
Principal components
Principal components analysis (PCA)
Regression adjustment
Residual control chart
Sample covariance matrix
Sample mean vector
Trajectory plots

EXERCISES

Note: Minitab's **Tsquared** multivariate control chart does not accept summary statistics, so many of these exercises were solved in Excel.

11.1.

The data shown in Table 11.E1 come from a production process with two observable quality characteristics: x^1 and x^2. The data are sample means of each quality characteristic, based on samples of size n -= 25. Assume that mean values of the quality characteristics and the covariance matrix were computed from 50 preliminary samples:

$$\bar{\bar{x}} = \begin{bmatrix} 55 \\ 30 \end{bmatrix} \quad S = \begin{bmatrix} 200 & 130 \\ 130 & 120 \end{bmatrix}$$

Construct a T^2 control chart using these data. Use the phase II limits.

■ TABLE 11E.1
Data for Exercise 11.1

Sample Number	x_1	x_2
1	58	32
2	60	33
3	50	27
4	54	31
5	63	38
6	53	30
7	42	20
8	55	31
9	46	25
10	50	29
11	49	27
12	57	30
13	58	33
14	75	45
15	55	27

$m = 50$ preliminary samples, $n = 25$ sample size, $p = 2$ characteristics. Let $\alpha = 0.001$.

$$UCL = \frac{p(m+1)(n-1)}{mn-m-p+1} F_{\alpha,p,mn-m-p+1}$$

$$= \frac{2(50+1)(25-1)}{50(25)-50-2+1} F_{0.001,2,1199} = (2448/1199)(6.948) = 14.186$$

LCL = 0

11.1. continued

(Note: Solution is in Student's Excel data set for Chapter 11)

p =	2	xdoublebars		S: Var-Covar matrix		S-1										
m =	50	55		200	130		0.0169	-0.0183								
n =	25	30		130	120		-0.0183	0.0282								
alpha =	0.001															
F =	6.9477															
Sample No.	1	2	3	4	5	6	7	8	9	10	11	12	13	14	15
xbar1	58	60	50	54	63	53	42	55	46	50	49	57	58	75	55
xbar2	32	33	27	31	38	30	20	31	25	29	27	30	33	45	27
diff1	3	5	-5	-1	8	-2	-13	0	-9	-5	-6	2	3	20	0
diff2	2	3	-3	1	8	0	-10	1	-5	-1	-3	0	3	15	-3
matrix calc	0.0451	0.1268	0.1268	0.0817	0.5408	0.0676	0.9127	0.0282	0.4254	0.2676	0.2028	0.0676	0.0761	2.1127	0.2535
t2 = n * calc	1.1268	3.1690	3.1690	2.0423	13.5211	1.6901	22.8169	0.7042	10.6338	6.6901	5.0704	1.6901	1.9014	52.8169	6.3380
UCL =	14.1851	14.1851	14.1851	14.1851	14.1851	14.1851	14.1851	14.1851	14.1851	14.1851	14.1851	14.1851	14.1851	14.1851	14.1851
LCL =	0	0	0	0	0	0	0	0	0	0	0	0	0	0	0
OOC?	In control	In control	In control	In control	In control	In control	Above UCL	In control	In control	In control	In control	In control	In control	Above UCL	In control

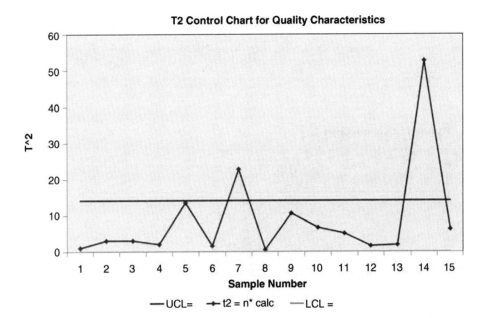

T2 Control Chart for Quality Characteristics

Legend: —UCL=   —◆— t2 = n* calc   —LCL =

Process is out of control at samples 7 and 14.

11.3.

Reconsider the situation in Exercise 11.1. Suppose that the sample mean vector and sample covariance matrix provided were the actual population parameters. What control limit would be appropriate for phase II for the control chart? Apply this limit to the data and discuss any differences in results that you find in comparison to the original choice of limits.

11.3. continued

Phase 2 T^2 control limits with $p = 2$ characteristics. Let $\alpha = 0.001$.

Since population parameters are known, the chi-square formula will be used for the upper control limit:

$$\text{UCL} = \chi^2_{\alpha,p} = \chi^2_{0.001,2} = 13.816$$

(Note: Solution is in Student's Excel data set for Chapter 11)

p =	2		xdoublebars		S: Var-Covar matrix		S-1		
m =	50		55		200	130	0.0169	-0.0183	
n =	25		30		130	120	-0.0183	0.0282	
alpha =	0.001								
X^2 =	13.8155								

Sample No.	1	2	3	4	5	6	7	8	9	10	11	12	13	14	15
xbar1	58	60	50	54	63	53	42	55	46	50	49	57	58	75	55
xbar2	32	33	27	31	38	30	20	31	25	29	27	30	33	45	27
diff1	3	5	-5	-1	8	-2	-13	0	-9	-5	-6	2	3	20	0
diff2	2	3	-3	1	8	0	-10	1	-5	-1	-3	0	3	15	-3
matrix calc	0.0451	0.1268	0.1268	0.0817	0.5408	0.0676	0.9127	0.0282	0.4254	0.2676	0.2028	0.0676	0.0761	2.1127	0.2535
t2 = n * calc	1.1268	3.1690	3.1690	2.0423	13.5211	1.6901	22.8169	0.7042	10.6338	6.6901	5.0704	1.6901	1.9014	52.8169	6.3380
UCL =	13.8155	13.8155	13.8155	13.8155	13.8155	13.8155	13.8155	13.8155	13.8155	13.8155	13.8155	13.8155	13.8155	13.8155	13.8155
LCL =	0	0	0	0	0	0	0	0	0	0	0	0	0	0	0
OOC?	In control	In control	In control	In control	In control	In control	Above UCL	In control	In control	In control	In control	In control	In control	Above UCL	In control

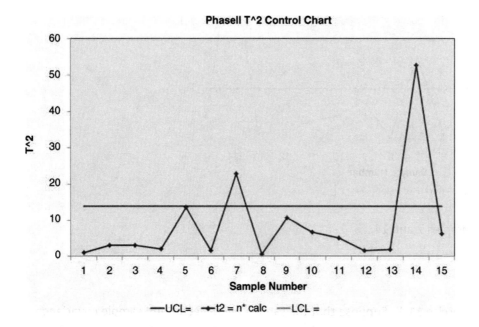

PhaseII T^2 Control Chart

(y-axis: T^2; x-axis: Sample Number)

—UCL= —t2 = n* calc —LCL =

Process is out of control at samples 7 and 14. Same results as for parameters estimated from samples.

11.5.

Consider a T^2 control chart for monitoring $p = 6$ quality characteristics. Suppose that the subgroup size is $n= 3$ and there are 30 preliminary samples available to estimate the sample covariance matrix.

$m = 30$ preliminary samples, $n = 3$ sample size, $p = 6$ characteristics, $\alpha = 0.005$

(a) Find the phase II control limits assuming that $\alpha = 0.005$.

Phase II limits:

$$UCL = \frac{p(m+1)(n-1)}{mn-m-p+1}F_{\alpha,p,mn-m-p+1}$$

$$= \frac{6(30+1)(3-1)}{30(3)-30-6+1}F_{0.005,6,55} = \left(\frac{372}{55}\right)(3.531) = 23.882$$

LCL = 0

(b) Compare the control limits from part (a) to the chi-square control limit. What is the magnitude of the difference in the two control limits?

chi-square limit: $UCL = \chi^2_{\alpha,p} = \chi^2_{0.005,6} = 18.548$

The Phase II UCL is almost 30% larger than the chi-square limit.

(c) How many preliminary samples would have to be taken to ensure that the exact phase II control limit is within 1% of the chi-square control limit?

Quality characteristics, $p = 6$. Samples size, $n = 3$. $\alpha = 0.005$. Find "m" such that exact Phase II limit is within 1% of chi-square limit, 1.01(18.548) = 18.733.

11.5.(c) continued

(Note: Solution is in Student's Excel data set for Chapter 11)

m	num	denom	F	UCL
30	372	55	3.531	23.8820
40	492	75	3.407	22.3527
50	612	95	3.338	21.5042
60	732	115	3.294	20.9648
70	852	135	3.263	20.5918
80	972	155	3.240	20.3185
90	1092	175	3.223	20.1096
100	1212	195	3.209	19.9448
...				
717	8616	1429	3.107	18.7336
718	8628	1431	3.107	18.7333
719	8640	1433	3.107	18.7331
720	8652	1435	3.107	18.7328
721	8664	1437	3.107	18.7325
722	8676	1439	3.107	18.7323

720 preliminary samples must be taken to ensure that the exact Phase II limit is within 1% of the chi-square limit.

11.7.

Consider a T^2 control chart for monitoring $p = 10$ quality characteristics. Suppose that the subgroup size is $n = 3$ and there are 25 preliminary samples available to estimate the sample covariance matrix.

$m = 25$ preliminary samples, $n = 3$ sample size, $p = 10$ characteristics, $\alpha = 0.005$

(a) Find the phase II control limits assuming that $\alpha = 0.005$.

Phase II UCL:

$$UCL = \frac{p(m+1)(n-1)}{mn-m-p+1} F_{\alpha,p,mn-m-p+1}$$

$$= \frac{10(25+1)(3-1)}{25(3)-25-10+1} F_{0.005,10,41} = \left(\frac{520}{41}\right)(3.101) = 39.326$$

(b) Compare the control limits from part (a) to the chi-square control limit. What is the magnitude of the difference in the two control limits?

chi-square UCL: $UCL = \chi^2_{\alpha,p} = \chi^2_{0.005,10} = 25.188$

The Phase II UCL is more than 55% larger than the chi-square limit.

11.7. continued

(c) How many preliminary samples would have to be taken to ensure that the exact phase II control limit is within 1% of the chi-square control limit?

Find "*m*" such that exact Phase II limit is within 1% of chi-square limit, 1.01(25.188) = 25.440.

(Note: Solution is in Student's Excel data set for Chapter 11)

m	num	denom	F	UCL
25	520	41	3.101	39.3261
35	720	61	2.897	34.1994
45	920	81	2.799	31.7953
55	1120	101	2.742	30.4023
65	1320	121	2.704	29.4939
75	1520	141	2.677	28.8548
85	1720	161	2.657	28.3808
95	1920	181	2.641	28.0153
105	2120	201	2.629	27.7248
986	19740	1963	2.530	25.4404
987	19760	1965	2.530	25.4402
988	19780	1967	2.530	25.4399
989	19800	1969	2.530	25.4397
990	19820	1971	2.530	25.4394

988 preliminary samples must be taken to ensure that the exact Phase II limit is within 1% of the chi-square limit.

11.9.
Consider a T^2 control chart for monitoring p = 10 quality characteristics. Suppose that the subgroup size is n = 3 and there are 25 preliminary samples available to estimate the sample covariance matrix. Calculate both the phase I and the phase II control limits (use α = 0.01).

p = 10 quality characteristics, n = 3 sample size, m = 25 preliminary samples. Assume α = 0.01.

Phase I UCL:

$$UCL = \frac{p(m-1)(n-1)}{mn-m-p+1}F_{\alpha,p,mn-m-p+1}$$

$$= \frac{10(25-1)(3-1)}{25(3)-25-10+1}F_{0.01,10,41}$$

$$= \left(\frac{480}{41}\right)(2.788)$$

$$= 32.638$$

11.9. continued

Phase II UCL:

$$UCL = \frac{p(m+1)(n-1)}{mn-m-p+1}F_{\alpha,p,mn-m-p+1}$$

$$= \frac{10(25+1)(3-1)}{25(3)-25-10+1}F_{0.01,10,41}$$

$$= \left(\frac{520}{41}\right)(2.788)$$

$$= 35.360$$

11.11.

Suppose that we have $p = 3$ quality characteristics, and in correlation form all three variables have variance unity and all pairwise correlation coefficients are 0.8. The in-control value of the process mean vector is $\mu' = [0, 0, 0]$.

(Note: Solutions are in Student's Excel data set for Chapter 11)

(a) Write out the covariance matrix Σ.

$$\Sigma = \begin{bmatrix} 1 & 0.8 & 0.8 \\ 0.8 & 1 & 0.8 \\ 0.8 & 0.8 & 1 \end{bmatrix}$$

(b) What is the chi-square control limit for the chart, assuming that $\alpha = 0.05$?

$$UCL = \chi^2_{\alpha,p} = \chi^2_{0.05,3} = 7.815$$

(c) Suppose that a sample of observations results in the standardized observation vector $y' = [1, 2, 0]$. Calculate the value of the T^2 statistic. Is an out-of-control signal generated?

$$T^2 = n(y-\mu)'\Sigma^{-1}(y-\mu)$$

$$= 1\left(\begin{bmatrix} 1 \\ 2 \\ 0 \end{bmatrix} - \begin{bmatrix} 0 \\ 0 \\ 0 \end{bmatrix}\right)'\begin{bmatrix} 1 & 0.8 & 0.8 \\ 0.8 & 1 & 0.8 \\ 0.8 & 0.8 & 1 \end{bmatrix}^{-1}\left(\begin{bmatrix} 1 \\ 2 \\ 0 \end{bmatrix} - \begin{bmatrix} 0 \\ 0 \\ 0 \end{bmatrix}\right)$$

$$= 11.154$$

Yes. Since $\left(T^2 = 11.154\right) > \left(UCL = 7.815\right)$, an out-of-control signal is generated.

11.11. continued

(d) Calculate the diagnostic quantities d_i = 1, 2, 3, from equation 11.22. Does this information assist in identifying which process variables have shifted?

$$\chi^2_{0.05,1} = 3.841$$

$$T^2_{(1)} = n\left(\mathbf{y}_{(1)} - \mathbf{\mu}_{(1)}\right)' \mathbf{\Sigma}^{-1}_{(1)}\left(\mathbf{y}_{(1)} - \mathbf{\mu}_{(1)}\right) = 1\left(\begin{bmatrix} 2 \\ 0 \end{bmatrix} - \begin{bmatrix} 0 \\ 0 \end{bmatrix}\right)' \begin{bmatrix} 1 & 0.8 \\ 0.8 & 1 \end{bmatrix}^{-1}\left(\begin{bmatrix} 2 \\ 0 \end{bmatrix} - \begin{bmatrix} 0 \\ 0 \end{bmatrix}\right) = 11.111$$

$$d_1 = T^2 - T^2_{(1)} = 11.154 - 11.111 = 0.043$$

$$T^2_{(2)} = 2.778; \quad d_2 = 8.376$$

$$T^2_{(3)} = 5.000; \quad d_3 = 6.154$$

Variables 2 and 3 should be investigated.

(e) Suppose that a sample of observations results in the standardized observation vector $\mathbf{y}'$ = [2, 2, 1]. Calculate the value of the T^2 statistic.

Since $(T^2 = 6.538) > (UCL = 7.815)$, an out-of-control signal is not generated.

(f) For the case in (e), calculate the diagnostic quantities d_i, i = 1, 2, 3 from equation 11.22. Does this information assist in identifying which process variables have shifted?

$$\chi^2_{0.05,1} = 3.841$$

$$T^2_{(1)} = 5.000; \quad d_1 = 1.538$$

$$T^2_{(2)} = 5.000; \quad d_2 = 1.538$$

$$T^2_{(3)} = 4.444; \quad d_3 = 2.094$$

Since an out-of-control signal was not generated in (e), it is not necessary to calculate the diagnostic quantities. This is confirmed since none of the d_i's exceeds the UCL.

11.13.

Consider the first three process variables in Table 11.5. Calculate an estimate of the sample covariance matrix using both estimators S_1 and S_2 discussed in Section 11.3.2.

■ **TABLE 11.5**
Cascade Process Data

Observation	x_1	x_2	x_3	x_4	x_5	x_6	x_7	x_8	x_9	y_1	Residuals	y_2
1	12.78	0.15	91	56	1.54	7.38	1.75	5.89	1.11	951.5	0.81498	87
2	14.97	0.1	90	49	1.54	7.14	1.71	5.91	1.109	952.2	−0.31685	88
3	15.43	0.07	90	41	1.47	7.33	1.64	5.92	1.104	952.3	−0.28369	86
4	14.95	0.12	89	43	1.54	7.21	1.93	5.71	1.103	951.8	−0.45924	89
5	16.17	0.1	83	42	1.67	7.23	1.86	5.63	1.103	952.3	−0.56512	86
6	17.25	0.07	84	54	1.49	7.15	1.68	5.8	1.099	952.2	−0.22592	91
7	16.57	0.12	89	61	1.64	7.23	1.82	5.88	1.096	950.2	−0.55431	99
8	19.31	0.08	99	60	1.46	7.74	1.69	6.13	1.092	950.5	−0.18874	100
9	18.75	0.04	99	52	1.89	7.57	2.02	6.27	1.084	950.6	0.15245	103
10	16.99	0.09	98	57	1.66	7.51	1.82	6.38	1.086	949.8	−0.33580	107

...

(Note: Solutions are in Student's Excel data set for Chapter 11)

$m = 40$

$$\bar{x}' = [15.339 \quad 0.104 \quad 88.125]; \quad S_1 = \begin{bmatrix} 4.440 & -0.016 & 5.395 \\ -0.016 & 0.001 & -0.014 \\ 5.395 & -0.014 & 27.599 \end{bmatrix}$$

$$V'V = \begin{bmatrix} 121.101 & -0.256 & 43.720 \\ -0.256 & 0.071 & 0.950 \\ 43.720 & 0.950 & 587.000 \end{bmatrix}; \quad S_2 = \begin{bmatrix} 1.553 & -0.003 & -0.561 \\ -0.003 & 0.001 & 0.012 \\ -0.561 & 0.012 & 7.526 \end{bmatrix}$$

11.15.

Suppose that there are $p = 4$ quality characteristics, and in correlation form all four variables have variance unity and all pairwise correlation coefficients are 0.75. The in-control value of the process mean vector is $\mu' = [0, 0, 0, 0]$, and we want to design an MEWMA control chart to provide good protection against a shift to a new mean vector of $y' = [1, 1, 1, 1]$. If an in-control ARL_0 of 200 is satisfactory, what value of λ and what upper control limit should be used? Approximately, what is the ARL_1 for detecting the shift in the mean vector?

11.15. continued

(Note: Solutions are in Student's Excel data set for Chapter 11)

p =	4			
mu' =	0	0	0	0

Sigma =	1	0.75	0.75	0.75
	0.75	1	0.75	0.75
	0.75	0.75	1	0.75
	0.75	0.75	0.75	1

Sigma-1 =	3.0769	-0.9231	-0.9231	-0.9231
	-0.9231	3.0769	-0.9231	-0.9231
	-0.9231	-0.9231	3.0769	-0.9231
	-0.9231	-0.9231	-0.9231	3.0769

y' =	1	1	1	1

y =	1
	1
	1
	1

y' Sigma-1 =	0.308	0.308	0.308	0.308

y' Sigma-1 y =	1.231

delta =	1.109

ARL0 =	200

From Table 10-3, select (lambda, H) pair that closely minimizes ARL1

delta =	1	1.5
lambda =	0.1	0.2
UCL = H =	12.73	13.87
ARL1 =	12.17	6.53

Select $\lambda = 0.1$ with an UCL = H = 12.73. This gives an ARL_1 between 7.22 and 12.17.

11.17.

Suppose that there are p = 2 quality characteristics, and in correlation form both variables have variance unity and the correlation coefficient is 0.8. The in-control value of the process mean vector is $\mu' = [0, 0]$, and we want to design an MEWMA control chart to provide good protection against a shift to a new mean vector of $y' = [1, 1]$. If an in-control ARL_0 of 200 is satisfactory, what value of λ and what upper control limit should be used? Approximately, what is the ARL_1 for detecting the shift in the mean vector?

11.17. continued

(Note: Solutions are in Student's Excel data set for Chapter 11)

p =	2							
mu' =	0	0						
Sigma =	1	0.8			Sigma-1 =	2.7778	-2.2222	
	0.8	1				-2.2222	2.7778	
y' =	1	1			y =		1	
							1	
y' Sigma-1 =	0.556	0.556						
y' Sigma-1 y =	1.111							
delta =	1.054							
ARL0 =	200							

From Table 10-3, select (lambda, H) pair that closely minimizes ARL1				
delta =	1	1	1.5	1.5
lambda =	0.1	0.2	0.2	0.3
UCL = H =	8.64	9.65	9.65	10.08
ARL1 =	10.15	10.20	5.49	5.48

Select $\lambda = 0.2$ with an UCL = H = 9.65. This gives an ARL_1 between 5.49 and 10.20.

11.19.

Consider the cascade process data in Table 11.5. In fitting regression models to both y_1 and y_2 you will find that not all of the process variables are required to obtain a satisfactory regression model for the output variables. Remove the nonsignificant variables from these equations and obtain subset regression models for both y_1 and y_2. Then construct individuals control charts for both sets of residuals. Compare them to the residual control charts in the text (Fig. 11.11) and from Exercise 11.18. Are there any substantial differences between the charts from the two different approaches to fitting the regression models?

Different approaches can be used to identify insignificant variables and reduce the number of variables in a regression model. This solution uses Minitab's "Best Subsets" functionality to identify the best-fitting model with as few variables as possible.

11.19. continued

Stat > Regression > Best Subsets

Best Subsets Regression: Tab11-5y1 versus Tab11-5x1, Tab11-5x2, ...
```
Response is Tab11-5y1
                                      T T T T T T T T T
                                      a a a a a a a a a
                                      b b b b b b b b b
                                      1 1 1 1 1 1 1 1 1
                                      0 0 0 0 0 0 0 0 0
                                      - - - - - - - - -
                                      5 5 5 5 5 5 5 5 5
                         Mallows      x x x x x x x x x
Vars  R-Sq  R-Sq(adj)    C-p       S  1 2 3 4 5 6 7 8 9
  1   43.1     41.6     52.9   1.3087               X
  1   31.3     29.5     71.3   1.4378       X
  2   62.6     60.5     24.5   1.0760   X           X
  2   55.0     52.5     36.4   1.1799       X       X
  3   67.5     64.7     18.9   1.0171   X   X       X
  3   66.8     64.0     19.9   1.0273   X         X X
  4   72.3     69.1     13.3   0.95201  X   X X     X
  4   72.1     68.9     13.6   0.95522  X   X     X X
  5   79.5     76.5      4.0   0.83020  X   X X   X X    ******
  5   73.8     69.9     13.0   0.93966  X   X     X X X
  6   79.9     76.2      5.5   0.83550  X   X X   X X X
  6   79.8     76.1      5.6   0.83693  X X X     X X
  7   80.3     76.0      6.8   0.83914  X   X X X X X X
  7   80.1     75.8      7.1   0.84292  X   X X X X X X
...
```

For output variable y_1, a regression model of input variables x_1, x_3, x_4, x_8, and x_9 maximize adjusted R^2 (minimize S) and minimize Mallow's C-p.

11.19. continued

Stat > Regression > Regression

Regression Analysis: Tab11-5y1 versus Tab11-5x1, Tab11-5x3, ...
```
The regression equation is
Tab11-5y1 = 819 + 0.431 Tab11-5x1 - 0.124 Tab11-5x3 - 0.0915 Tab11-5x4
            + 2.64 Tab11-5x8 + 115 Tab11-5x9


Predictor       Coef   SE Coef        T       P
Constant      818.80     29.14    28.10   0.000
Tab11-5x1    0.43080   0.08113     5.31   0.000
Tab11-5x3   -0.12396   0.03530    -3.51   0.001
Tab11-5x4   -0.09146   0.02438    -3.75   0.001
Tab11-5x8     2.6367    0.7604     3.47   0.001
Tab11-5x9     114.81     23.65     4.85   0.000


S = 0.830201   R-Sq = 79.5%   R-Sq(adj) = 76.5%


Analysis of Variance
Source           DF        SS       MS       F       P
Regression        5    90.990   18.198   26.40   0.000
Residual Error   34    23.434    0.689
Total            39   114.424
```

Stat > Control Charts > Variables Charts for Individuals > Individuals

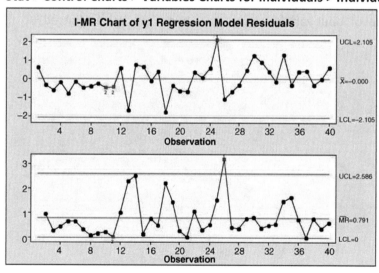

11.19. continued

Test Results for I Chart of Ex11.19Res1
```
TEST 1. One point more than 3.00 standard deviations from center line.
Test Failed at points:   25
TEST 2. 9 points in a row on same side of center line.
Test Failed at points:   10, 11
```

Test Results for MR Chart of Ex11.19Res1
```
TEST 1. One point more than 3.00 standard deviations from center line.
Test Failed at points:   26
TEST 2. 9 points in a row on same side of center line.
Test Failed at points:   11
```

Stat > Regression > Best Subsets

Best Subsets Regression: Tab11-5y2 versus Tab11-5x1, Tab11-5x2, ...
```
Response is Tab11-5y2
                                            T T T T T T T T T
                                            a a a a a a a a a
                                            b b b b b b b b b
                                            1 1 1 1 1 1 1 1 1
                                            0 0 0 0 0 0 0 0 0
                                            - - - - - - - - -
                                            5 5 5 5 5 5 5 5 5
                              Mallows        x x x x x x x x x
Vars  R-Sq  R-Sq(adj)   C-p       S      1 2 3 4 5 6 7 8 9
   1  36.1      34.4   24.0   4.6816        X
   1  35.8      34.1   24.2   4.6921                      X
   2  55.1      52.7    8.1   3.9751                    X X
   2  50.7      48.1   12.2   4.1665        X           X
   3  61.6      58.4    4.0   3.7288        X X         X
   3  59.8      56.4    5.7   3.8160        X           X X
   4  64.9      60.9    2.9   3.6147    X   X           X X
   4  64.4      60.4    3.4   3.6387        X X         X X
   5  67.7      62.9    2.3   3.5208    X   X X         X X    ******
   5  65.2      60.1    4.7   3.6526    X   X       X   X X
   6  67.8      62.0    4.2   3.5660    X   X X       X X X
   6  67.8      61.9    4.3   3.5684    X   X X X       X X
   7  67.9      60.9    6.1   3.6149    X X X X       X X X
   7  67.8      60.8    6.2   3.6200    X   X X X     X X X
...
```

For output variable y_2, a regression model of input variables x_1, x_3, x_4, x_8, and x_9 maximize adjusted R^2 (minimize S) and minimize Mallow's C-p.

11.19. continued

Stat > Regression > Regression
Regression Analysis: Tab11-5y2 versus Tab11-5x1, Tab11-5x3, ...
```
The regression equation is
Tab11-5y2 = 244 - 0.633 Tab11-5x1 + 0.454 Tab11-5x3 + 0.176 Tab11-5x4
            + 11.2 Tab11-5x8 - 236 Tab11-5x9


Predictor       Coef   SE Coef       T       P
Constant       244.4     123.6    1.98   0.056
Tab11-5x1    -0.6329    0.3441   -1.84   0.075
Tab11-5x3     0.4540    0.1497    3.03   0.005
Tab11-5x4     0.1758    0.1034    1.70   0.098
Tab11-5x8     11.175     3.225    3.47   0.001
Tab11-5x9     -235.7     100.3   -2.35   0.025


S = 3.52081   R-Sq = 67.7%   R-Sq(adj) = 62.9%


Analysis of Variance
Source          DF        SS       MS       F       P
Regression       5    882.03   176.41   14.23   0.000
Residual Error  34    421.47    12.40
Total           39   1303.50
```

Stat > Control Charts > Variables Charts for Individuals > Individuals

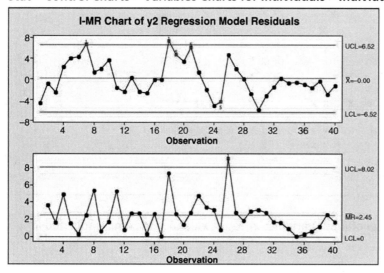

11.19. continued

```
Test Results for I Chart of Ex11.19Res2
TEST 1. One point more than 3.00 standard deviations from center line.
Test Failed at points:   7, 18
TEST 5. 2 out of 3 points more than 2 standard deviations from center line (on
     one side of CL).
Test Failed at points:   19, 21, 25
TEST 6. 4 out of 5 points more than 1 standard deviation from center line (on
     one side of CL).
Test Failed at points:   7, 21

Test Results for MR Chart of Ex11.19Res2
TEST 1. One point more than 3.00 standard deviations from center line.
Test Failed at points:   26
```

For response y_1, there is not a significant difference between control charts for residuals from either the full regression model (Figure 11.10, no out-of-control observations) and the subset regression model (observation 25 is OOC).

For response y_2, there is not a significant difference between control charts for residuals from either the full regression model (Exercise 11.18, observations 7 and 18 are OOC) and the subset regression model (observations 7 and 18 are OOC).

11.21.

Consider the $p = 4$ process variables in Table 11.6. After applying the PCA procedure to the first 20 observations data (see Table 11.7), suppose that the first three principal components are retained.

■ TABLE 11.6
Chemical Process Data

Observation	x_1	x_2	x_3	x_4	z_1	z_2
			Original Data			
1	10	20.7	13.6	15.5	0.291681	−0.6034
2	10.5	19.9	18.1	14.8	0.294281	0.491533
3	9.7	20	16.1	16.5	0.197337	0.640937
4	9.8	20.2	19.1	17.1	0.839022	1.469579
5	11.7	21.5	19.8	18.3	3.204876	0.879172
6	11	20.9	10.3	13.8	0.203271	−2.29514
7	8.7	18.8	16.9	16.8	−0.99211	1.670464
8	9.5	19.3	15.3	12.2	−1.70241	−0.36089
9	10.1	19.4	16.2	15.8	−0.14246	0.560808
10	9.5	19.6	13.6	14.5	−0.99498	−0.31493

11.21. continued

(a) Obtain the principal component scores. (Hint: Remember that you must work in standardized variables.)

Stat > Multivariate > Principal Components

To work in standardized variables in Minitab, select Correlation Matrix.

To obtain principal component scores, select Storage and enter columns for Scores.

Principal Component Analysis: Tab11-6x1, Tab11-6x2, Tab11-6x3, Tab11-6x4

```
Eigenanalysis of the Correlation Matrix

Eigenvalue   2.3181   1.0118   0.6088   0.0613
Proportion   0.580    0.253    0.152    0.015
Cumulative   0.580    0.832    0.985    1.000

Variable      PC1      PC2      PC3      PC4
Tab11-6x1   0.594   -0.334    0.257    0.685
Tab11-6x2   0.607   -0.330    0.083   -0.718
Tab11-6x3   0.286    0.794    0.534   -0.061
Tab11-6x4   0.444    0.387   -0.801    0.104
```

Principal Component Scores:

z1	z2	z3
0.29168	-0.60340	0.02496
0.29428	0.49153	1.23823
0.19734	0.64094	-0.20787
0.83902	1.46958	0.03929
3.20488	0.87917	0.12420
0.20327	-2.29514	0.62545
-0.99211	1.67046	-0.58815
-1.70241	-0.36089	1.82157
-0.14246	0.56081	0.23100
-0.99498	-0.31493	0.33164
0.94470	0.50471	0.17975
-1.21950	-0.09129	-1.11787
2.60867	-0.42176	-1.19166
-0.12378	-0.08767	-0.19592
-1.10423	1.47259	0.01299
-0.27825	-0.94763	-1.31445
-2.65608	0.13529	-0.11243
2.36528	-1.30494	0.32285
0.41131	-0.21893	0.64479
-2.14662	-1.17849	-0.86838

11.21. continued

(b) Construct an appropriate set of pairwise plots of the principal component scores.

Graph > Matrix Plot > Simple Matrix of Plots

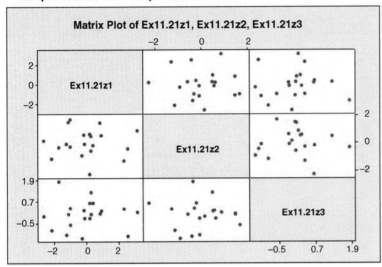

(c) Calculate the principal component scores for the last 10 observations. Plot the scores on the charts from part (b) and interpret the results.

Note: Principal component scores for new observations were calculated in Excel. See solution in the Student's Excel data set for Chapter 11.

Graph > Matrix Plot > Matrix of Plots with Groups

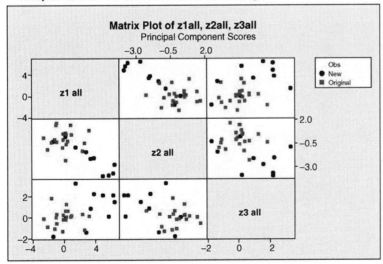

Although a few new points are within area defined by the original points, the majority of new observations are clearly different from the original observations.

Chapter 12

Engineering Process Control and SPC

LEARNING OBJECTIVES

After completing this chapter you should be able to:

1. Explain the origins of process monitoring and process adjustment
2. Explain the different statistical frameworks of SPC and EPC
3. Explain how an integral controller works
4. Understand how EPC transfers variability from the process output into a manipulatable variable
5. Set up and use a manual adjustment chart
6. Understand the basis of the bounded adjustment chart
7. Understand the basis of proportional integral (PI) and proportional integral derivative (PID) controllers
8. Know how to combine SPC and EPC by applying a control chart to the output quality characteristic

IMPORTANT TERMS AND CONCEPTS

Automatic process control (APC)

Bounded adjustment chart

Engineering process control (EPC)

Exponentially weighted moving average (EWMA)

Integral control

Integrating SPC and EPC

Manipulatable process variable

Process adjustment versus process monitoring

Process gain

Proportional integral (PI) control

Proportional integral derivative (PID) control

Setpoint

Statistical process control (SPC)

Statistical process monitoring

EXERCISES

Note: Minitab does not include functionality for some of the methods presented in Chapter 12; however Excel can be used to solve the exercises.

12.3.

Consider the data in Table 12.1. Construct a bounded adjustment chart using $\lambda = 0.4$ and $L = 10$. Compare the performance of this chart to the one in Table 12.1 and Fig. 12.12.

■ TABLE 12.1
Chemical Process Data for the Bounded Adjustment Chart in Figure 12.12

Observation	Original Process Output	Adjusted Process Output	EWMA	Adjustment	Cumulative Adjustment or Setpoint
1	0	0	0		0
2	16	16	3.200		0
3	24	24	7.360		0
4	29	29	11.688	-7.250	-7.250
5	34	26.750	5.350		-7.250
6	24	16.750	7.630		-7.250
7	31	23.750	10.854	-5.938	-13.188
8	26	12.812	2.562		-13.188
9	38	24.812	7.013		-13.188
10	29	15.812	8.773		-13.188
11	25	11.812	9.381		-13.188
12	26	12.812	10.067	-3.203	-16.391
13	23	6.609	1.322		-16.391
14	34	17.609	4.579		-16.391
15	24	7.609	5.185		-16.391
16	14	-2.391	3.670		-16.391
17	41	24.609	7.858		-16.391
18	36	19.609	10.208	-4.904	-21.293
19	29	7.707	1.541		-21.293
20	13	-8.293	-0.425		-21.293

...

12.3. continued

(Note: Solution is in Student's Excel data set for Chapter 12)

Target yt =	0						
lambda =	0.4						
L =	10						
g =	0.8						

Obs	Orig_out	Orig_Nt	Adj_out_t	EWMA_t	|EWMA_t|>L?	Adj_Obs_t+1	Cum_Adj
1	0	0					
2	16	16	16	6.400	no	0	0
3	24	8	24	13.440	yes	-12	-12
4	29	5	17	6.800	no	0	-12
5	34	5	22	12.880	yes	-11	-23
6	24	-10	1	0.400	no	0	-23
7	31	7	8	3.440	no	0	-23
8	26	-5	3	3.264	no	0	-23
9	38	12	15	7.958	no	0	-23
45	22	9	10.5	3.564	no	0	-11.5
46	-9	-31	-20.5	-6.061	no	0	-11.5
47	3	12	-8.5	-7.037	no	0	-11.5
48	12	9	0.5	-4.022	no	0	-11.5
49	3	-9	-8.5	-5.813	no	0	-11.5
50	12	9	0.5	-3.288	no	0	-11.5
SS =	21468		5610.25				
Average =	17.24		0.91				

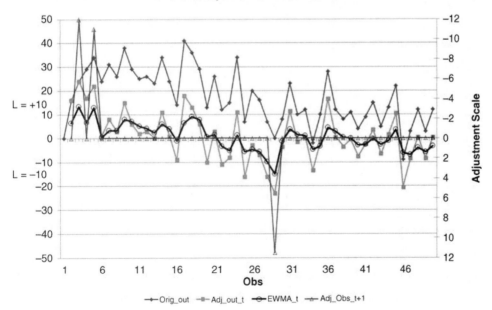

Bounded Adjustment Chart for Ex 12-3

The chart with $\lambda = 0.4$ exhibits less variability, but is further from target on average than for the chart with $\lambda = 0.3$.

12.5.

The Variogram. Consider the variance of observations that are m periods apart; that is, $V_m = V(y_{t+m} - y_t)$. A graph of V_m/V_1 versus m is called a variogram. It is a nice way to check a data series for nonstationary (drifting mean) behavior. If a data series is completely uncorrelated (white noise) the variogram will always produce a plot that stays near unity. If the data series is autocorrelated but stationary, the plot of the variogram will increase for a while, but as m increases the plot of V_m/V_1 will gradually stabilize and not increase any further. The plot of V_m/V_1 versus m will increase without bound for nonstationary data. Apply this technique to the data in Table 12.1. Is there an indication of nonstationary behavior? Calculate the sample autocorrelation function for the data. Compare the interpretation of both graphs.

(Note: Solution is in Student's Excel data set for Chapter 12)

t	Yt / m =>	1	2	3	4	5	6	7	8	9
1	0									
2	16	16								
3	24	8	24							
4	29	5	13	29						
5	34	5	10	18	34					
6	24	-10	-5	0	8	24				
7	31	7	-3	2	7	15	31			
8	26	-5	2	-8	-3	2	10	26		
9	38	12	7	14	4	9	14	22	38	
10	29	-9	3	-2	5	-5	0	5	13	29

...

	Var_m =	147.11	175.72	147.47	179.02	136.60	151.39	162.43	201.53	138.70
Var_m/Var_1 =		1.000	1.195	1.002	1.217	0.929	1.029	1.104	1.370	0.943

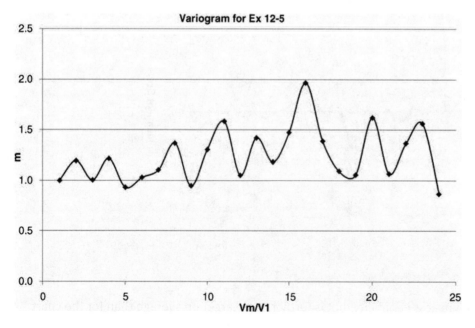

12.5. continued

MTB : Chap12.mtw : Yt

Stat > Time Series > Autocorrelation Function

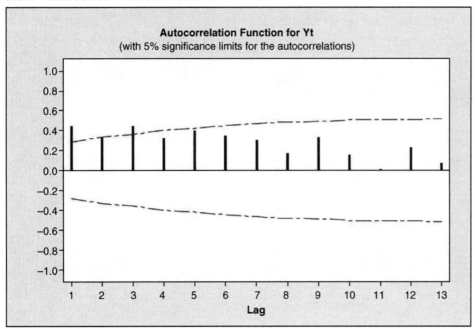

Autocorrelation Function: Yt

```
Lag        ACF       T     LBQ
  1   0.440855   3.12   10.31
  2   0.334961   2.01   16.39
  3   0.440819   2.45   27.14
  4   0.316478   1.58   32.80
  5   0.389094   1.85   41.55
  6   0.345327   1.54   48.59
  7   0.299822   1.28   54.03
  8   0.164698   0.68   55.71
  9   0.325056   1.33   62.41
 10   0.149321   0.59   63.86
 11   0.012158   0.05   63.87
 12   0.228540   0.90   67.44
 13   0.066173   0.26   67.75
```

Variogram appears to be increasing, so the observations are correlated and there may be some mild indication of nonstationary behavior. The slow decline in the sample ACF also indicates the data are correlated and potentially nonstationary.

12.7.

Use the data in Exercise 12.6 to construct a bounded adjustment chart. Use $\lambda = 0.2$ and set $L = 12$. How does the bounded adjustment chart perform relative to the integral control adjustment procedure in part (a) of Exercise 12.6?

(Note: Solutions are in Student's Excel data set for Chapter 12)

Target yt =	200
lambda =	0.2
L =	12
g =	1.2

Obs, t	Orig_out	Orig_Nt	Adj_out_t	EWMA_t	|EWMA_t|>L?	Adj_Obs_t+1	Cum_Adj
1	215.8	0					
2	195.8	-20	196	-0.840	no	0.0	0.0
3	191.3	-4.5	191.300	-2.412	no	0.0	0.000
4	185.3	-6	185.300	-4.870	no	0.000	0.000
5	216.0	30.7	216.000	-0.696	no	0.000	0.000
6	176.9	-39.1	176.900	-5.177	no	0.000	0.000
7	176.0	-0.9	176.000	-8.941	no	0.000	0.000
8	162.6	-13.4	162.600	-14.633	yes	6.233	6.233
9	187.5	24.9	193.733	-1.253	no	0.000	6.233
10	180.5	-7	186.733	-3.656	no	0.000	6.233
11	174.5	-6	180.733	-6.778	no	0.000	6.233
12	151.6	-22.9	157.833	-13.856	yes	7.028	13.261
42	152.3	8.6	185.919	-8.184	no	0.000	33.619
43	111.3	-41	144.919	-17.563	yes	9.180	42.799
44	143.6	32.3	186.399	-2.720	no	0.000	42.799
45	129.9	-13.7	172.699	-7.636	no	0.000	42.799
46	122.9	-7	165.699	-12.969	yes	5.717	48.516
47	126.2	3.3	174.716	-5.057	no	0.000	48.516
48	133.2	7	181.716	-7.702	no	0.000	48.516
49	145.0	11.8	193.516	-7.459	no	0.000	48.516
50	129.5	-15.5	178.016	-10.364	no	0.000	48.516

SS =	1,323,872		1,632,265
Average =	161.304		182.051
Variance =	467.8		172.7

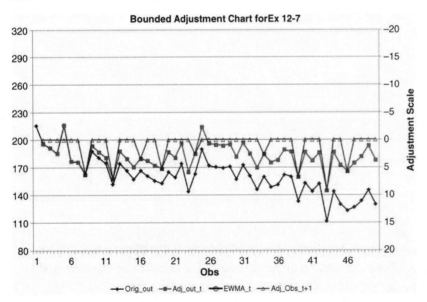

Bounded Adjustment Chart forEx 12-7

Behavior of the bounded adjustment control scheme is similar to both integral control schemes ($\lambda = 0.2$ and $\lambda = 0.4$).

12.9.

Consider the observations in Table 12E.2. The target value for this process is 50.

▪ TABLE 12E.2
Process Data for Exercise 12.9

Observation, t	y_t	Observation, t	y_t
1	50	26	43
2	58	27	39
3	54	28	32
4	45	29	37
5	56	30	44
6	56	31	52
7	66	32	42
8	55	33	47
9	69	34	33
10	56	35	49
11	63	36	34
12	54	37	40

13	67	38	27
14	55	39	29
15	56	40	35
16	65	41	27
17	65	42	33
18	61	43	25
19	57	44	21
20	61	45	16
21	64	46	24
22	43	47	18
23	44	48	20
24	45	49	23
25	39	50	26

(Note: Solutions are in Student's Excel data set for Chapter 12)

(a) Set up an integral controller for this process. Assume that the gain for the adjustment variable is $g = 1.6$ and assume that $\lambda = 0.2$ in the EWMA forecasting procedure will provide adequate one-step-ahead predictions.

T =	50				
lambda =	0.2				
g =	1.6				
Obs	Orig_out	Orig_Nt	Adj_out_t	Adj_Obs_t+1	Cum_Adj
1	50				
2	58	8.0	58.0	-1.0	-1.0
3	54	-4.0	53.0	-0.4	-1.4
4	45	-9.0	43.6	0.8	-0.6
5	56	11.0	55.4	-0.7	-1.3
6	56	0.0	54.7	-0.6	-1.8
7	66	10.0	64.2	-1.8	-3.6
8	55	-11.0	51.4	-0.2	-3.8
9	69	14.0	65.2	-1.9	-5.7
10	56	-13.0	50.3	0.0	-5.7
11	63	7.0	57.3	-0.9	-6.6
45	16	-5.0	32.3	2.2	18.5
46	24	8.0	42.5	0.9	19.4
47	18	-6.0	37.4	1.6	21.0
48	20	2.0	41.0	1.1	22.1
49	23	3.0	45.1	0.6	22.7
50	26	3.0	48.7	0.2	22.9
	Unadjusted		Adjusted		
SS =	109,520		108,629		
Average =	44.4		46.262		
Variance =	223.51		78.32		

12.9.(a) continued

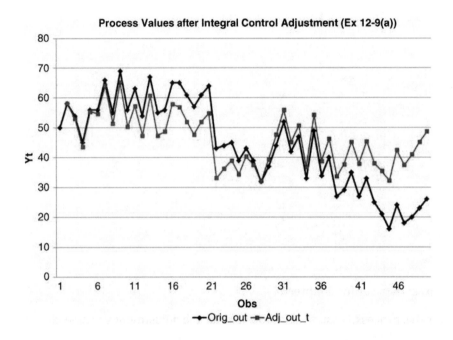

(b) How much reduction in variability around the target does the integral controller achieve?

Based on the reduction in variance, there is significant reduction in variability with use of integral control scheme.

(c) Rework parts (a) and (b) assuming that $\lambda = 0.4$. What change does this make in the variability around the target in comparison to that achieved with $\lambda = 0.2$?

12.9.(c) continued

Obs	Orig_out	Orig_Nt	Adj_out_t	Adj_Obs_t+1	Cum_Adj
T =	50				
lambda =	0.4				
g =	1.6				
1	50				
2	58	8.0	58.0	-2.0	-2.0
3	54	-4.0	52.0	-0.5	-2.5
4	45	-9.0	42.5	1.9	-0.6
5	56	11.0	55.4	-1.3	-2.0
6	56	0.0	54.0	-1.0	-3.0
7	66	10.0	63.0	-3.3	-6.2
8	55	-11.0	48.8	0.3	-5.9
9	69	14.0	63.1	-3.3	-9.2
10	56	-13.0	46.8	0.8	-8.4
11	63	7.0	54.6	-1.2	-9.5
45	16	-5.0	37.4	3.1	24.6
46	24	8.0	48.6	0.4	24.9
47	18	-6.0	42.9	1.8	26.7
48	20	2.0	46.7	0.8	27.5
49	23	3.0	50.5	-0.1	27.4
50	26	3.0	53.4	-0.8	26.5
SS =	109,520		114,819		
Average =	44.4		47.833		
Variance =	223.51		56.40		

There is a slight reduction in variability with use of $\lambda = 0.4$, as compared to $\lambda = 0.2$, with a process average slightly closer to the target of 50.

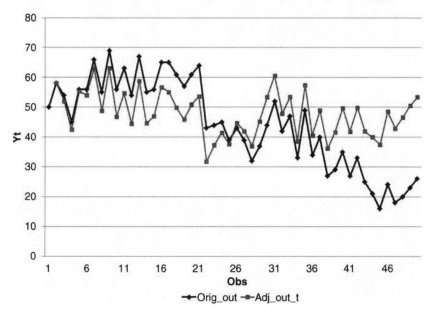

Process Values after Integral Control Adjustment (Ex 12-9(c))

Chapter 13

Factorial and Fractional Factorial Experiments for Process Design and Improvement

LEARNING OBJECTIVES

After completing this chapter you should be able to:

1. Explain how designed experiments can be used to improve product design and improve process performance
2. Explain how designed experiments can be used to reduce the cycle time required to develop new products and processes
3. Understand how main effects and interactions of factors can be estimated
4. Understand the factorial design concept
5. Know how to use the analysis of variance (ANOVA) to analyze data from factorial designs
6. Know how residuals are used for model adequacy checking for factorial designs
7. Know how to use the 2^k system of factorial designs
8. Know how to construct and interpret contour plots and response surface plots
9. Know how to add center points to a 2^k factorial design to test for curvature and provide an estimate of pure experimental error
10. Understand how the blocking principal can be used in a factorial design to eliminate the effects of a nuisance factor
11. Know how to use the 2^{k-p} system of fractional factorial designs

IMPORTANT TERMS AND CONCEPTS

2^k factorial designs	Guidelines for planning experiments
2^{k-p} fractional factorial designs	Interaction
Aliasing	Main effect of a factor
Analysis of variance (ANOVA)	Normal probability plot of effects
Analysis procedure for factorial designs	Orthogonal design
Blocking	Pre-experimental planning
Center points in a 2k and 2k p factorial designs	Projection of 2^k and 2^{k-p} factorial designs
Completely randomized design	Regression model representation of experimental results
Confounding	Residual analysis
Contour plot	Residuals

Controllable process variables	Resolution of a fractional factorial design
Curvature in the response function	Response surface
Defining relation for a fractional factorial design	Screening Experiments
Factorial design	Sequential experimentation
Fractional factorial design	Sparsity of effects principle
Generators for a fractional factorial design	Two-factor interaction

EXERCISES

Note: To analyze an experiment in Minitab, the initial experimental layout must be created in Minitab or defined by the user. The Excel data sets contain only the data given in the textbook; therefore some information required by Minitab is not included. Detailed Minitab instructions are provided for Exercise 13.3 to define and create designs. The remaining exercises are worked in a similar manner, and only the solutions are provided.

13.1.

The following output was obtained from a computer program that performed a two-factor ANOVA on a factorial experiment.

Source	SS	DF	MS	F	P
A	0.322	1			
B	80.554		40.2771		
Interaction					
Error	108.327	12			
Total	231.551	17			

(a) Fill in the blanks in the ANOVA table. You can use bounds on the P-values.

(Note: The Instructors' Excel data set for Chapter 13 has the solution.)

Source	SS	DF	MS	F	P
A	0.322	1	0.322	0.036	0.853
B	80.554	2	40.2771	4.462	0.036
Interaction	42.348	2	21.174	2.346	0.138
Error	108.327	12	9.02725		
Total	231.551	17			

13.1. continued

(b) How many levels were used for factor B?

$DF_B = b - 1 \rightarrow b = DF_B + 1 = 2 + 1 = 3$

(c) How many replicates of the experiment were performed?

$DF_{Total} = abn - 1 \rightarrow n = (DF_{Total} + 1)/ab = (17 + 1) / (2 \times 3) = 3$

(d) What conclusions would you draw about this experiment?

The *P*-value for Factor B is very small (less than $\alpha = 0.05$), indicating that it significantly affects the response. Additionally, the large *P*-values for Factor A and the Interaction term indicate that they are not significant.

13.3.
An article in Industrial Quality Control (1956, pp. 5–8) describes an experiment to investigate the effect of glass type and phosphor type on the brightness of a television tube. The response measured is the current necessary (in microamps) to obtain a specified brightness level. The data are shown here. Analyze the data and draw conclusions.

■ TABLE 13E.1
Data for Exercise 13.1

Glass Type	Phosphor Type		
	1	2	3
1	280	300	290
	290	310	285
	285	295	290
2	230	260	220
	235	240	225
	240	235	230

This experiment is three replicates of a factorial design in two factors—two levels of glass type and three levels of phosphor type—to investigate brightness. Enter the data into the Minitab worksheet using the first three columns: one column for glass type, one column for phosphor type, and one column for brightness. This is how the Excel data set is structured. Since the experiment layout was not created in Minitab, the design must be defined before the results can be analyzed.

13.3. continued

After entering the data in Minitab, select **Stat > DOE > Factorial > Define Custom Factorial Design**. Select the two factors (Glass Type and Phosphor Type), then for this exercise, check "**General full factorial**". The dialog box should look like this:

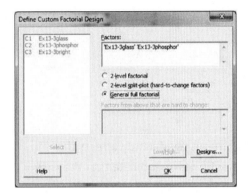

Next, select "**Designs**". For this exercise, no information is provided on standard order, run order, point type, or blocks, so leave the selections as below, and click "**OK**" twice.

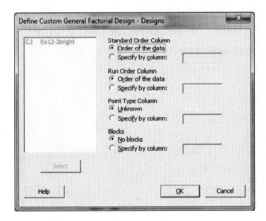

Note that Minitab added four new columns (4 through 7) to the worksheet. DO NOT insert or delete columns between columns 1 through 7. Minitab recognizes these contiguous seven columns as a designed experiment; inserting or deleting columns will cause the design layout to become corrupt.

13.3. continued

↓	C1	C2	C3	C4	C5	C6	C7
	Ex13-3glass	Ex13-3phosphor	Ex13-3bright	StdOrder	RunOrder	Blocks	CenterPt
1	1	1	280	1	1	1	1
2	1	1	290	2	2	1	1
3	1	1	285	3	3	1	1
4	1	2	300	4	4	1	1
5	1	2	310	5	5	1	1
6	1	2	295	6	6	1	1
7	1	3	290	7	7	1	1
8	1	3	285	8	8	1	1
9	1	3	290	9	9	1	1
10	2	1	230	10	10	1	1
11	2	1	235	11	11	1	1
12	2	1	240	12	12	1	1
13	2	2	260	13	13	1	1
14	2	2	240	14	14	1	1
15	2	2	235	15	15	1	1
16	2	3	220	16	16	1	1
17	2	3	225	17	17	1	1
18	2	3	230	18	18	1	1

The design and data are in the Minitab worksheet **Ex13-3.MTW**.

Select **Stat > DOE > Factorial > Analyze Factorial Design**. Select the response (**Brightness**), then click on "**Terms**", verify that the selected terms are Glass Type, Phosphor Type, and their interaction, click "**OK**". Leave "**Regular**" residuals selected. Click on "**Graphs**", select "**Residuals Plots : Four in one**". The option to plot residuals versus variables is for continuous factor levels; since the factor levels in this experiment are categorical, do not select this option. Click "**OK**". Click on "**Storage**", select "**Fits**" and "**Residuals**", and click "**OK**" twice.

General Linear Model: Ex13-3bright versus Ex13-3glass, Ex13-3phosphor

```
Factor          Type   Levels  Values
Ex13-3glass     fixed     2    1, 2
Ex13-3phosphor  fixed     3    1, 2, 3

Analysis of Variance for Ex13-3bright, using Adjusted SS for Tests
Source                       DF    Seq SS    Adj SS    Adj MS       F      P
Ex13-3glass                   1   14450.0   14450.0   14450.0  273.79  0.000
Ex13-3phosphor                2     933.3     933.3     466.7    8.84  0.004
Ex13-3glass*Ex13-3phosphor    2     133.3     133.3      66.7    1.26  0.318
Error                        12     633.3     633.3      52.8
Total                        17   16150.0

S = 7.26483   R-Sq = 96.08%   R-Sq(adj) = 94.44%
```

13.3. continued

No indication of significant interaction (*P*-value is greater than 0.10). Glass type (A) and phosphor type (B) significantly affect television tube brightness (*P*-values are less than 0.10).

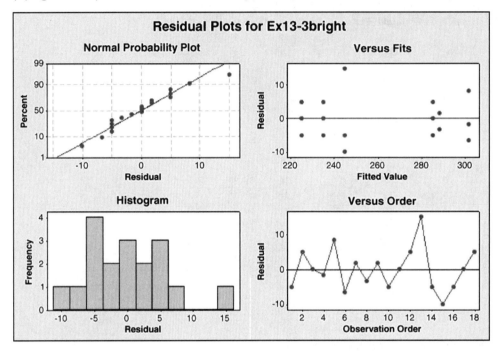

Visual examination of residuals on the normal probability plot, histogram, and versus fitted values reveals no problems. The plot of residuals versus observation order is not meaningful since no order was provided with the data. If the model were re-fit with only Glass Type and Phosphor Type, the residuals should be re-examined.

To plot residuals versus the two factors, select **Graph > Individual Value Plot > One Y with Groups**. Select the column with stored residuals (**RESI1**) as the **Graph variable** and select one of the factors (Glass Type or Phosphor Type) as the **Categorical variable for grouping**. Click on "**Scale**", select the "**Reference Lines**" tab, and enter "**0**" for the Y axis, then click "**OK**" twice.

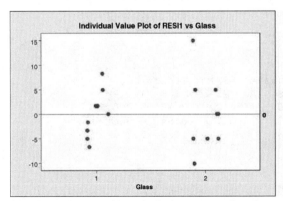

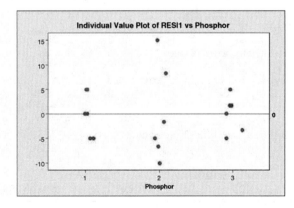

13.3. continued

Note that the plot points are "jittered" about the factor levels. To remove the jitter, select the graph to make it active then: **Editor > Select Item > Individual Symbols** and then **Editor > Edit Individual Symbols > Jitter** and de-select **Add jitter to direction**.

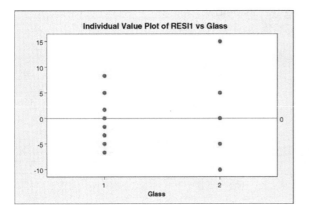
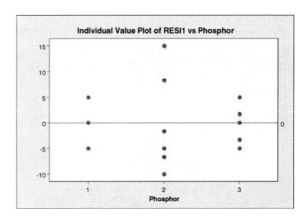

Variability appears to be the same for both glass types; however, there appears to be more variability in results with phosphor type 2.

Select **Stat > DOE > Factorial > Factorial Plots**. Select **"Interaction Plot"** and click on **"Setup"**, select the response (Brightness) and both factors (Glass Type and Phosphor Type), and click **"OK"** twice.

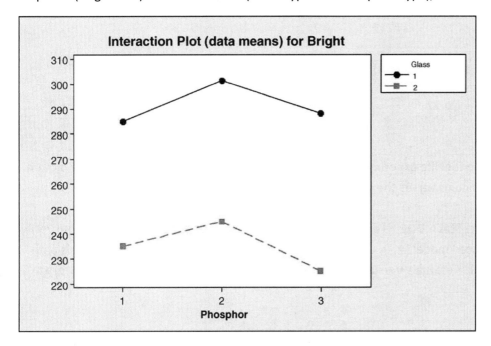

The absence of a significant interaction is evident in the parallelism of the two lines. Final selected combination of glass type and phosphor type depends on the desired brightness level.

13.3. continued

Alternate Solution: This exercise may also be solved using Minitab's ANOVA functionality instead of its DOE functionality. The DOE functionality was selected to illustrate the approach that will be used for most of the remaining exercises. To obtain results which match the output in the textbook's Table 13.5, select **Stat > ANOVA > Two-Way**, and complete the dialog box as below.

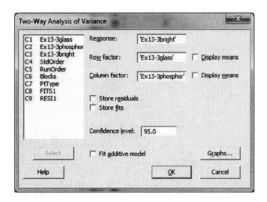

Two-way ANOVA: Ex13-3bright versus Ex13-3glass, Ex13-3phosphor

Source	DF	SS	MS	F	P
Ex13-3glass	1	14450.0	14450.0	273.79	0.000
Ex13-3phosphor	2	933.3	466.7	8.84	0.004
Interaction	2	133.3	66.7	1.26	0.318
Error	12	633.3	52.8		
Total	17	16150.0			

S = 7.265   R-Sq = 96.08%   R-Sq(adj) = 94.44%

13.5.

Find the residuals from the tool life experiment in Exercise 13.4. Construct a normal probability plot of the residuals. Plot the residuals versus the predicted values. Comment on the plots.

To find the residuals, select **Stat > DOE > Factorial > Analyze Factorial Design**. Select "**Terms**" and verify that all terms for the reduced model (A, B, C, AC) are included. Select "**Graphs**", and for residuals plots choose "**Normal plot**" and "**Residuals versus fits**". To save residuals to the worksheet, select "**Storage**" and choose "**Residuals**".

13.5. continued

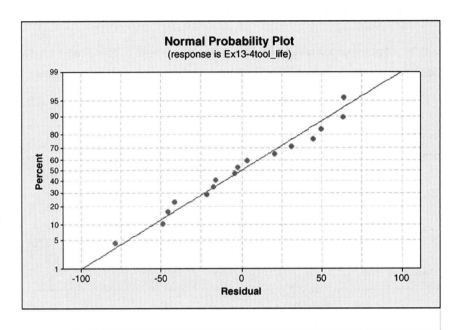

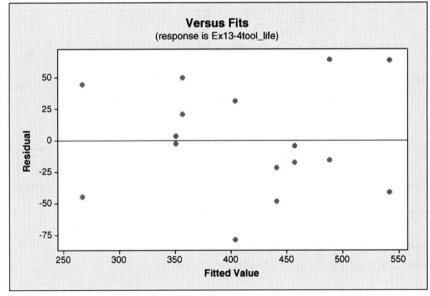

Normal probability plot of residuals indicates that the normality assumption is reasonable. Residuals versus fitted values plot shows that the equal variance assumption across the prediction range is reasonable.

13.7.

Consider the experiment in Exercise 13.6. Plot the residuals against the levels of factors A, B, C, and D. Also construct a normal probability plot of the residuals comment on these plots.

To find the residuals, select **Stat > DOE > Factorial > Analyze Factorial Design**. Select "**Terms**" and verify that all terms for the reduced model are included. Select "**Graphs**", choose "**Normal plot**" of residuals and "**Residuals versus variables**", and then select the variables.

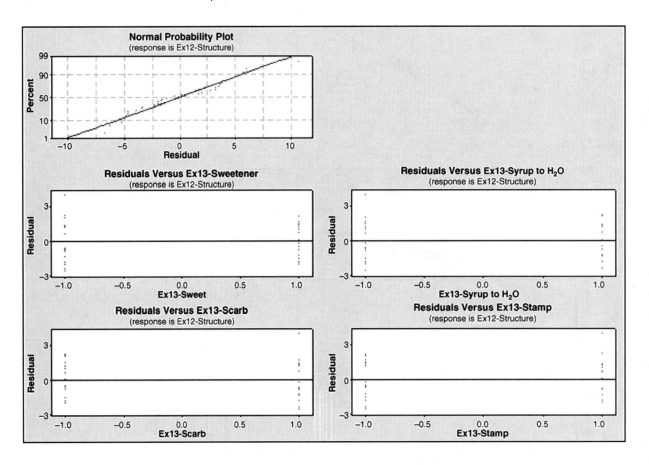

There appears to be a slight indication of inequality of variance for sweetener and syrup ratio, as well as a slight indication of an outlier. This is not serious enough to warrant concern.

13.9.

Suppose that only the data from replicate I in Exercise 13.6 were available. Analyze the data and draw appropriate conclusions.

(Note: The design and data are in the Minitab worksheet **Ex13-9.MTW**.)

Factorial Fit: Total Score versus Sweetener, Syrup to Water, ...

Estimated Effects and Coefficients for Total Score (coded units)

Term	Effect	Coef
Constant		183.625
Sweetener	-10.500	-5.250
Syrup to Water	-0.250	-0.125
Carbonation	0.750	0.375
Temperature	5.500	2.750
Sweetener*Syrup to Water	4.000	2.000
Sweetener*Carbonation	1.000	0.500
Sweetener*Temperature	-6.250	-3.125
Syrup to Water*Carbonation	-1.750	-0.875
Syrup to Water*Temperature	-3.000	-1.500
Carbonation*Temperature	1.000	0.500
Sweetener*Syrup to Water*Carbonation	-7.500	-3.750
Sweetener*Syrup to Water*Temperature	4.250	2.125
Sweetener*Carbonation*Temperature	0.250	0.125
Syrup to Water*Carbonation*Temperature	-2.500	-1.250
Sweetener*Syrup to Water*Carbonation*Temperature	3.750	1.875

...

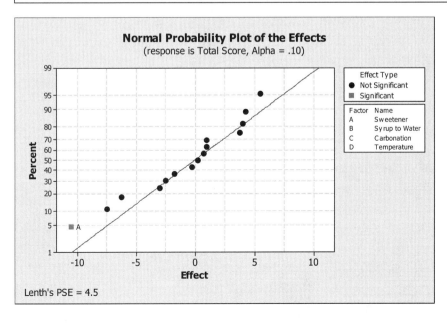

Lenth's PSE = 4.5

13.9. continued

From visual examination of the normal probability plot of effects, only factor A (sweetener) is significant. Re-fit and analyze the reduced model.

Factorial Fit: Total Score versus Sweetener

```
Estimated Effects and Coefficients for Total Score (coded units)
Term            Effect      Coef   SE Coef       T       P
Constant                 183.625     1.865   98.48   0.000
Sweetener     -10.500    -5.250     1.865   -2.82   0.014


S = 7.45822    R-Sq = 36.15%    R-Sq(adj) = 31.59%


Analysis of Variance for Total Score (coded units)
Source           DF    Seq SS    Adj SS   Adj MS      F       P
Main Effects      1    441.00   441.000   441.00   7.93   0.014
Residual Error   14    778.75   778.750    55.63
  Pure Error     14    778.75   778.750    55.63
Total            15   1219.75
```

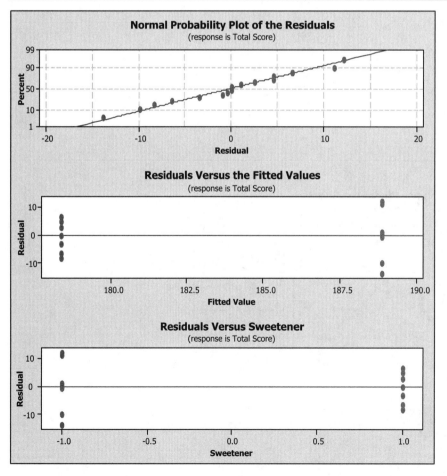

13.9. continued

There appears to be a slight indication of inequality of variance for sweetener, as well as in the predicted values. This is not serious enough to warrant concern.

13.11.

Show how a 2^5 experiment could be set up in two blocks of 16 runs each. Specifically, which runs would be made in each block?

A 2^5 design in two blocks will lose the ABCDE interaction to blocks.

Block 1		Block 2	
(1)	ae	a	e
ab	be	b	abe
ac	ce	c	ace
bc	abce	abc	bce
ad	de	d	ade
bd	abde	abd	bde
cd	acde	acd	cde
abcd	bcde	bcd	abcde

13.13.

An article in *Industrial and Engineering Chemistry* ("More on Planning Experiments to Increase Research Efficiency," 1970, pp. 60–65) uses a 2^{5-2} design to investigate the effect of A = condensation temperature, B = amount of material 1, C = solvent volume, D = condensation time, and E = amount of material 2, on yield. The results obtained are as follows:

$$e = 23.2 \qquad cd = 23.8$$
$$ab = 15.5 \qquad ace = 23.4$$
$$ad = 16.9 \qquad bde = 16.8$$
$$bc = 16.2 \qquad abcde = 18.1$$

(Note: The design and data are in the Minitab worksheet **Ex13-13.MTW**.)

(a) Verify that the design generators used were $I = ACE$ and $I = BDE$.

(b) Write down the complete defining relation and the aliases from this design.

13.13.(b) continued

Note: Minitab does not allow you to select the design generators $I = ACE$ and $I = BDE$. After creating a 2^{5-2} with the principal fraction and default generators, reverse the signs for factor D for the first four standard order runs to obtain the design in the exercise. (This is also shown in the Minitab worksheet.) Do NOT insert any columns between the design columns. Another option in Minitab would be to define the experiment using **Stat > DOE > Factorial > Define Custom Factorial Design.**

Select **Stat > DOE > Factorial > Analyze Factorial Design.** Since there is only one replicate of the experiment, select "**Terms**" and verify that all main effects and two-factor interaction effects are selected.

Factorial Fit: Yield versus Temp, Matl1, Vol, Time, Matl2

```
Estimated Effects and Coefficients for Yield (coded units)
Term          Effect    Coef
Constant                19.238
Temp          -1.525   -0.763
Matl1         -5.175   -2.587
Vol            2.275    1.138
Time          -0.675   -0.337
Matl2          2.275    1.138
Temp*Matl1     1.825    0.912
Temp*Time     -1.275   -0.638
...
Alias Structure
I + Temp*Vol*Matl2 + Matl1*Time*Matl2 + Temp*Matl1*Vol*Time
Temp + Vol*Matl2 + Matl1*Vol*Time + Temp*Matl1*Time*Matl2
Matl1 + Time*Matl2 + Temp*Vol*Time + Temp*Matl1*Vol*Matl2
Vol + Temp*Matl2 + Temp*Matl1*Time + Matl1*Vol*Time*Matl2
Time + Matl1*Matl2 + Temp*Matl1*Vol + Temp*Vol*Time*Matl2
Matl2 + Temp*Vol + Matl1*Time + Temp*Matl1*Vol*Time*Matl2
Temp*Matl1 + Vol*Time + Temp*Time*Matl2 + Matl1*Vol*Matl2
Temp*Time + Matl1*Vol + Temp*Matl1*Matl2 + Vol*Time*Matl2
```

From the Alias Structure shown in the Session Window, the complete defining relation is:
I = ACE = BDE = ABCD.

The aliases are:

A*I = A*ACE = A*BDE = A*ABCD ⇒ A = CE = ABDE = BCD

B*I = B*ACE = B*BDE = B*ABCD ⇒ B = ABCE = DE = ACD

C*I = C*ACE = C*BDE = C*ABCD ⇒ C = AE = BCDE = ABD

...AB*I = AB*ACE = AB*BDE = AB*ABCD ⇒ AB = BCE = ADE = CD

The remaining aliases are calculated in a similar fashion.

13.13. continued

(c) Estimate the main effects.

[A] = A + CE + BCD + ABDE
       = ¼ (−23.2 + 15.5 + 16.9 − 16.2 − 23.8 + 23.4 − 16.8 + 18.1) = ¼ (−6.1) = −1.525

[AB] = AB + BCE + ADE + CD
       = ¼ (+23.2 +15.5 − 16.9 -16.2 +23.8 − 23.4 − 16.8 + 18.1) = ¼ (7.3) = 1.825

This are the same effect estimates provided in the Minitab output above. The other main effects and interaction effects are calculated in the same way.

(d) Prepare an analysis of variance table. Verity that the *AB* and *AD* interactions are available to use as error.

Select **Stat > DOE > Factorial > Analyze Factorial Design.** Since there is only one replicate of the experiment, select "**Terms**" and verify that all main effects and two-factor interaction effects are selected. Then select "**Graphs**", choose the normal effects plot, and set alpha to 0.10.

```
Factorial Fit: Yield versus Temp, Matl1, Vol, Time, Matl2
...
Analysis of Variance for yield (coded units)
Source              DF   Seq SS   Adj SS   Adj MS   F   P
Main Effects         5   79.826   79.826   15.965   *   *
2-Way Interactions   2    9.913    9.913    4.956   *   *
Residual Error       0       *        *        *
Total                7   89.739
...
```

13.13.(d) continued

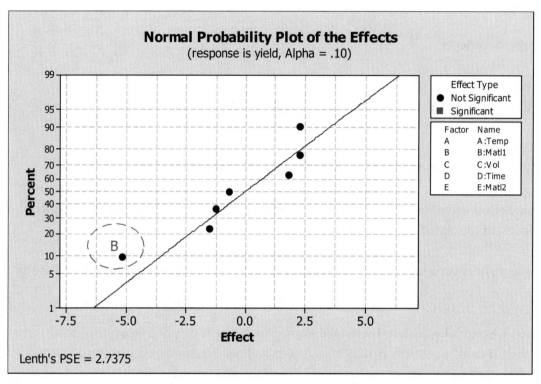

Although none of the effects is significant at 0.10, main effect B (amount of material 1) is more than twice as large as the 2nd largest effect (absolute values) and falls far from a line passing through the remaining points. Re-fit a reduced model containing only the B main effect, and pool the remaining terms to estimate error.

Select **Stat > DOE > Factorial > Analyze Factorial Design.** Select "**Terms**" and select "**B**". Then select "**Graphs**", and select the "**Normal plot**" and "**Residuals versus fits**" residual plots.

```
Factorial Fit: Yield versus Matl1
Estimated Effects and Coefficients for Yield (coded units)

Term      Effect    Coef   SE Coef     T      P
Constant           19.238  0.8682   22.16  0.000
Matl1     -5.175  -2.587   0.8682   -2.98  0.025

S = 2.45552      PRESS = 64.3156
R-Sq = 59.69%    R-Sq(pred) = 28.33%    R-Sq(adj) = 52.97%

Analysis of Variance for Yield (coded units)
Source          DF  Seq SS  Adj SS  Adj MS    F      P
Main Effects     1   53.56   53.56  53.561  8.88  0.025
  Matl1          1   53.56   53.56  53.561  8.88  0.025
Residual Error   6   36.18   36.18   6.030
  Pure Error     6   36.18   36.18   6.030
Total            7   89.74
  ...
```

13.13. continued

(e) Plot the residuals versus the fitted values. Also construct a normal probability plot of the residuals.
Comment on the results.

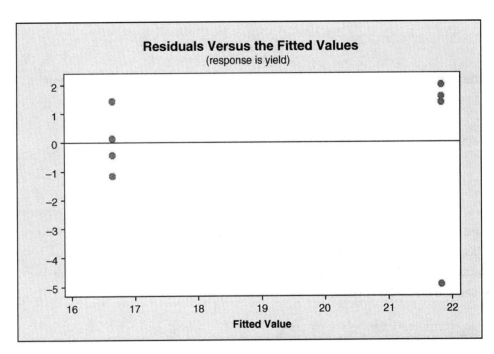

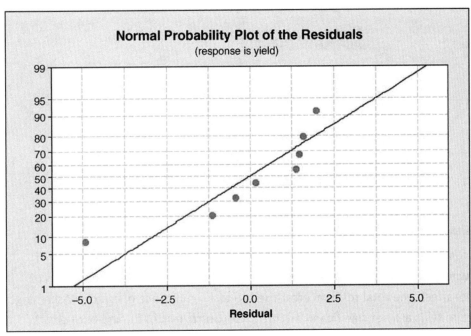

13.13.(e) continued

Residual plots indicate a potential outlier. The run should be investigated for any issues which occurred while running the experiment. If no issues can be identified, it may be necessary to make additional experimental runs.

13.15.

Reconsider the data in Exercise 13.14. Suppose that four center points were added to this experiment. The molecular weights at the center point are 90, 87, 86, and 93.

(Note: The design and data are in the Minitab worksheet **Ex13-15.MTW.**)

(a) Analyze the data as you did in Exercise 13.14, but include a test for curvature.

Select **Stat > DOE > Factorial > Analyze Factorial Design**. Select "**Terms**" and verify that all main effects and two-factor interactions are selected. Also, DO NOT include the center points in the model (uncheck the default selection). This will ensure that if both lack of fit and curvature are not significant, the main and interaction effects are tested for significance against the correct residual error (lack of fit + curvature + pure error). See the dialog box below.

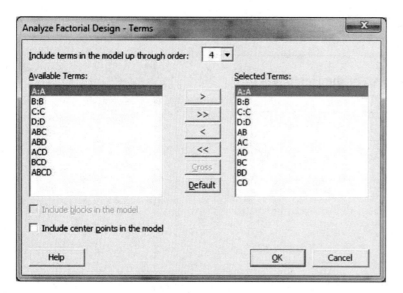

To summarize Minitab's functionality, curvature is always tested against pure error and lack of fit (if available), regardless of whether center points are included in the model. The inclusion/exclusion of center points in the model affects the total residual error used to test significance of effects. Assuming that lack of fit and curvature tests are not significant, all three (curvature, lack of fit, and pure error) should be included in the residual mean square.

13.15.(a) continued

When looking at results in the ANOVA table, the first test to consider is the "lack of fit" test, which is a test of significance for terms not included in the model (in this exercise, the three-factor and four-factor interactions). If lack of fit is significant, the model is not correctly specified, and some terms need to be added to the model.

If lack of fit is not significant, the next test to consider is the "curvature" test, which is a test of significance for the pure quadratic terms. If this test is significant, no further statistical analysis should be performed because the model is inadequate.

If tests for both lack of fit and curvature are not significant, then it is reasonable to pool the curvature, pure error, and lack of fit (if available) and use this as the basis for testing for significant effects. (In Minitab, this is accomplished by not including center points in the model.)

Factorial Fit: Ex13-15mole_wt versus A, B, C, D

```
Estimated Effects and Coefficients for Ex13-15mole_wt (coded units)
Term       Effect    Coef   SE Coef       T       P
Constant            84.800   0.8521   99.52   0.000
A          -3.750  -1.875    0.9527   -1.97   0.081
B           1.000   0.500    0.9527    0.52   0.612
C          -3.000  -1.500    0.9527   -1.57   0.150
D          -0.750  -0.375    0.9527   -0.39   0.703
A*B         2.250   1.125    0.9527    1.18   0.268
A*C        -0.250  -0.125    0.9527   -0.13   0.898
A*D         0.500   0.250    0.9527    0.26   0.799
B*C        -2.000  -1.000    0.9527   -1.05   0.321
B*D         0.250   0.125    0.9527    0.13   0.898
C*D         0.750   0.375    0.9527    0.39   0.703

...


Analysis of Variance for Ex13-15mole_wt (coded units)
Source             DF    Seq SS    Adj SS    Adj MS       F       P
Main Effects        4    98.500    98.500   24.6250    1.70   0.234
2-Way Interactions  6    40.000    40.000    6.6667    0.46   0.822
Residual Error      9   130.700   130.700   14.5222
  Curvature         1    88.200    88.200   88.2000   16.60   0.004
  Lack of Fit       5    12.500    12.500    2.5000    0.25   0.915
  Pure Error        3    30.000    30.000   10.0000
Total              19   269.200
...
```

(b) The test for curvature is significant (*P*-value = 0.004). Although one could pick a "winning combination" from the experimental runs, a better strategy is to add runs that would enable estimation of the quadratic effects. This approach to sequential experimentation is presented in Chapter 14.

13.17.

A 2 4-1 design has been used to investigate the effect of four factors on the resistivity of a silicon wafer. The data from this experiment are shown in Table 13E.4.

■ **TABLE 13E.4**
Resistivity Experiment for Exercise 13.17.

Run	A	B	C	D	Resistivity
1	−	−	−	−	33.2
2	+	−	−	+	4.6
3	−	+	−	+	31.2
4	+	+	−	−	9.6
5	−	−	+	+	40.6
6	+	−	+	−	162.4
7	−	+	+	−	39.4
8	+	+	+	+	158.6
9	0	0	0	0	63.4
10	0	0	0	0	62.6
11	0	0	0	0	58.7
12	0	0	0	0	60.9

Enter the factor levels and resist data into a Minitab worksheet, including a column indicating whether a run is a center point run (1 = not center point, 0 = center point). Then define the experiment using **Stat > DOE > Factorial > Define Custom Factorial Design**. The design and data are in the Minitab worksheet **Ex13-17.MTW**.

(a) Estimate the factor effects. Plot the effect estimates on a normal probability scale.

Select **Stat > DOE > Factorial > Analyze Factorial Design**. Select "**Terms**" and verify that all main effects and two-factor interactions are selected. Also, DO NOT include the center points in the model (uncheck the default selection). Then select "**Graphs**", choose the normal effects plot, and set alpha to 0.10.

13.17.(a) continued

Factorial Fit: Ex13-17Resist versus Ex13-17A, Ex13-17B, ...

```
Estimated Effects and Coefficients for Ex13-17Resist (coded units)
Term        Effect     Coef   SE Coef       T       P
Constant             60.433    0.6223   97.12   0.000
A           47.700   23.850    0.7621   31.29   0.000 *
B           -0.500   -0.250    0.7621   -0.33   0.759
C           80.600   40.300    0.7621   52.88   0.000 *
D           -2.400   -1.200    0.7621   -1.57   0.190
A*B          1.100    0.550    0.7621    0.72   0.510
A*C         72.800   36.400    0.7621   47.76   0.000 *
A*D         -2.000   -1.000    0.7621   -1.31   0.260

...

Analysis of Variance for Ex13-17Resist (coded units)
Source              DF    Seq SS    Adj SS    Adj MS        F       P
Main Effects         4   17555.3   17555.3   4388.83   944.51   0.000
2-Way Interactions   3   10610.1   10610.1   3536.70   761.13   0.000
Residual Error       4      18.6      18.6      4.65
  Curvature          1       5.6       5.6      5.61     1.30   0.338
  Pure Error         3      13.0      13.0      4.33
Total               11   28184.0
```

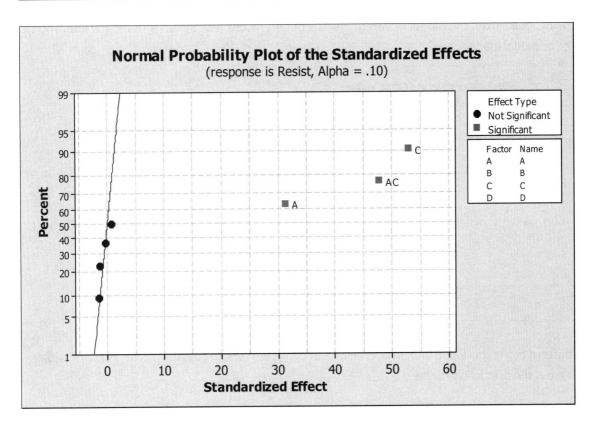

Normal Probability Plot of the Standardized Effects
(response is Resist, Alpha = .10)

Effect Type
● Not Significant
■ Significant

Factor	Name
A	A
B	B
C	C
D	D

13.17. continued

(b) Identify a tentative model for this process. Fit the model and test for curvature.

Examining the normal probability plot of effects, the main effects A and C and their two-factor interaction (AC) are significant. Re-fit and analyze a reduced model containing A, C, and AC.

Factorial Fit: Ex13-17Resist versus Ex13-17A, Ex13-17C

Estimated Effects and Coefficients for Ex13-17Resist (coded units)

Term	Effect	Coef	SE Coef	T	P	
Constant		60.43	0.6537	92.44	0.000	
A	47.70	23.85	0.8007	29.79	0.000	*
C	80.60	40.30	0.8007	50.33	0.000	*
A*C	72.80	36.40	0.8007	45.46	0.000	*

...

Analysis of Variance for Ex13-17Resist (coded units)

Source	DF	Seq SS	Adj SS	Adj MS	F	P
Main Effects	2	17543.3	17543.3	8771.6	1710.43	0.000
2-Way Interactions	1	10599.7	10599.7	10599.7	2066.89	0.000
Residual Error	8	41.0	41.0	5.1		
Curvature	1	5.6	5.6	5.6	1.11	0.327
Pure Error	7	35.4	35.4	5.1		
Total	11	28184.0				

Curvature is not significant (*P*-value = 0.327), so continue with analysis.

(c) Plot the residuals from the model in part (b) versus the predicted resistivity. Is there any indication on this plot of model inadequacy?

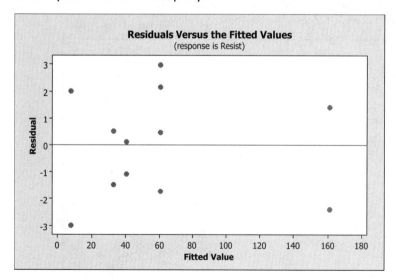

A funnel pattern at the low value and an overall lack of consistent width suggest a problem with equal variance across the prediction range.

13.17. continued

(d) Construct a normal probability plot of the residuals. Is there any reason to doubt the validity of the normality assumption?

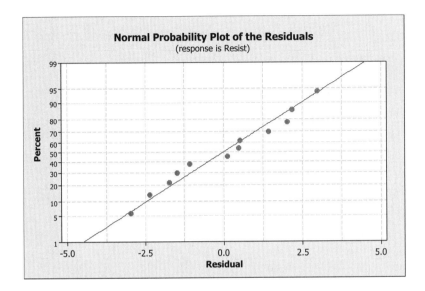

The normal probability plot of residuals is satisfactory.

The concern with variance in the predicted resistivity indicates that a data transformation may be needed.

Chapter 14

Process Optimization with Designed Experiments

LEARNING OBJECTIVES

After completing this chapter you should be able to:

1. Know how to use the response surface approach to optimizing processes
2. Know how to apply the method of steepest ascent
3. Know how to analyze a second-order response surface model
4. Know how to determine optimum operating conditions for a process
5. Know how to set up and conduct an experiment using a central composite design
6. Understand the difference between controllable process variables and noise variables
7. Understand the advantages of a combined array design for a process robustness study
8. Know how to use the response model approach to conduct a process robustness study
9. Know how evolutionary operation (EVOP) is used to maintain a process that is subject to drift near its optimum operating conditions

IMPORTANT TERMS AND CONCEPTS

Central composite design

Combined array design

Contour plot

Controllable variable

Crossed array design

Evolutionary operation (EVOP)

EVOP cycle

EVOP phase

First-order model

Inner array design

Method of steepest ascent

Noise variable

Outer array design

Path of steepest ascent

Process robustness study

Response model

Response surface

Response surface methodology (RSM)

Robust parameter design (RPD)

Rotatable design

Second-order model

Sequential experimentation

Taylor series

Transmission of error

Exercises

Note: To analyze an experiment in Minitab, the initial experimental layout must be created in Minitab or defined by the user. The Excel data sets contain only the data given in the textbook; therefore some information required by Minitab is not included. The Minitab instructions provided for the factorial designs in Chapter 13 are similar to those for response surface designs in this Chapter.

14.3.

Consider the first-order model $\hat{y}_t = 75 + 10x_1 + 6x_2$

(a) Sketch the contours of constant predicted response over the range $-1 \leq x_i \leq +1$, $i = 1, 2$.

In Minitab, designate columns for x_1 and x_2, generate a grid of values from -1 to $+1$ (**Calc > Make Mesh Data**), then calculate y.

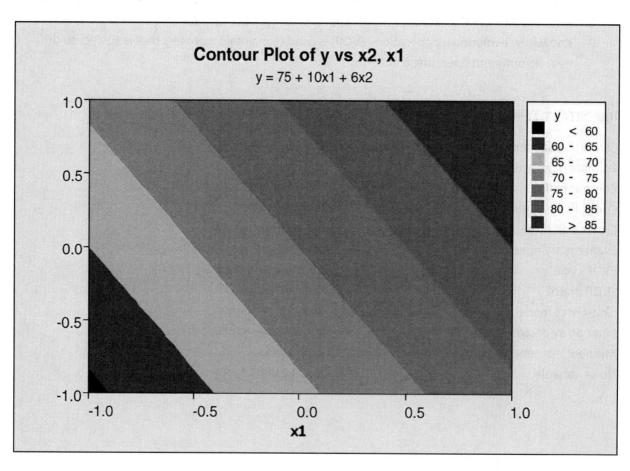

14.3. continued

(b) Find the direction of steepest ascent.

$$\hat{y} = 75 + 10x_1 + 6x_2 \quad -1 \le x_1 \le 1; 1 \le x_2 \le 1$$

$$\frac{x_2}{x_1} = \frac{6}{10} = 0.6$$

$$\Delta x_1 = 1$$

$$\Delta x_2 = 0.6$$

14.5.

An experiment was run to study the effect of two factors, time and temperature, on the inorganic impurity levels in paper pulp. The results of this experiment are shown in Table 14E.1.

■ TABLE 14E.1
The Experiment for
Exercise 14.5

x_1	x_2	y
−1	−1	210
1	−1	95
−1	1	218
1	1	100
−1.5	0	225
1.5	0	50
0	−1.5	175
0	1.5	180
0	0	145
0	0	175
0	0	158
0	0	166

(a) What type of experimental design has been used in this study? Is the design rotatable?

This design is a CCD with $k = 2$ and $\alpha = 1.5$. The design is not rotatable.

14.5. continued

(b) Fit a quadratic model to the response, using the method of least squares.

(Note: The design and data are in the Minitab worksheet **Ex14-5.MTW**.)

Select **Stat > DOE > Response Surface > Analyze Response Surface Design.** Select **"Terms"** and verify that all main effects, two-factor interactions, and quadratic terms are selected.

```
Response Surface Regression: Ex14-5y versus Ex14-5x1, Ex14-5x2

The analysis was done using coded units.

Estimated Regression Coefficients for Ex14-5y
Term                    Coef   SE Coef        T      P
Constant             160.868     4.555   35.314  0.000
Ex14-5x1             -87.441     4.704  -18.590  0.000
Ex14-5x2               3.618     4.704    0.769  0.471
Ex14-5x1*Ex14-5x1    -24.423     7.461   -3.273  0.017
Ex14-5x2*Ex14-5x2     15.577     7.461    2.088  0.082
Ex14-5x1*Ex14-5x2     -1.688    10.285   -0.164  0.875
...
Analysis of Variance for Ex14-5y
Source              DF    Seq SS    Adj SS    Adj MS       F      P
Regression           5   30583.4   30583.4    6116.7   73.18  0.000
  Linear             2   28934.2   28934.2   14467.1  173.09  0.000
  Square             2    1647.0    1647.0     823.5    9.85  0.013
  Interaction        1       2.3       2.3       2.3    0.03  0.875
Residual Error       6     501.5     501.5      83.6
  Lack-of-Fit        3      15.5      15.5       5.2    0.03  0.991
  Pure Error         3     486.0     486.0     162.0
Total               11   31084.9
...
Estimated Regression Coefficients for Ex14-5y using data in uncoded units
Term                     Coef
Constant              160.868
Ex14-5x1             -58.2941
Ex14-5x2              2.41176
Ex14-5x1*Ex14-5x1    -10.8546
Ex14-5x2*Ex14-5x2     6.92314
Ex14-5x1*Ex14-5x2    -0.750000
```

14.5. continued

(c) Construct the fitted impurity response surface. What values of x_1 and x_2 would you recommend if you wanted to minimize the impurity level?

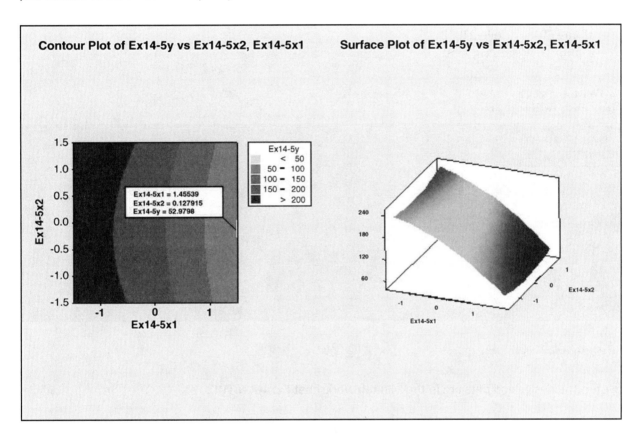

From visual examination of the contour and surface plots, it appears that minimum purity can be achieved by setting x_1 (time) = +1.5 and letting x_2 (temperature) range from –1.5 to + 1.5. The range for x_2 agrees with the ANOVA results indicating that it is statistically insignificant (P-value = 0.471). The level for temperature could be established based on other considerations, such as cost. A flag is planted at one option on the contour plot above.

(d) Suppose that $x_1 = \dfrac{\text{temp} - 750}{50}$ $x_2 = \dfrac{\text{time} - 30}{15}$ where temperature is in °C and time is in hours. Find the optimum operating conditions in terms of the natural variables temperature and time.

$\text{Temp} = 50x_1 + 750 = 50(+1.50) + 750 = 825$

$\text{Time} = 15x_2 + 30 = 15(-0.22) + 30 = 26.7$

14.7.

An article in *Rubber Chemistry and Technology* (Vol. 47, 1974, pp. 825–836) describes an experiment that studies the relationship of the Mooney viscosity of rubber to several variables, including silica filler (parts per hundred) and oil filler (parts per hundred). Some of the data from this experiment are shown in Table 14E.2, where

$$x_1 = \frac{silica - 60}{15} \qquad x_2 = \frac{oil - 21}{1.5}$$

■ **TABLE 14E.2**
The Viscosity Experiment for Exercise 14.7

Coded Levels		
x_1	x_2	y
−1	−1	13.71
1	−1	14.15
−1	1	12.87
1	1	13.53
−1.4	0	12.99
1.4	0	13.89
0	−1.4	14.16
0	1.4	12.90
0	0	13.75
0	0	13.66
0	0	13.86
0	0	13.63
0	0	13.74

(Note: The design and data are in the Minitab worksheet **Ex14-7.MTW**.)

(a) What type of experimental design has been used? Is it rotatable?

The design is a CCD with $k = 2$ and $\alpha = 1.4$. The design is rotatable.

(b) Fit a quadratic model to these data. What values of x_1 and x_2 will maximize the Mooney viscosity?

Since the standard order is provided, one approach to solving this exercise in Minitab is to create a two-factor response surface design in Minitab, then enter the data.

Select **Stat > DOE > Response Surface > Create Response Surface Design**. Leave the design type as a 2-factor, central composite design. Select "**Designs**", highlight the design with five center points (13 runs), and enter a custom alpha value of exactly 1.4 (the rotatable design is $\alpha = 1.41421$). Under "**Options**" uncheck Randomize runs to leave the runs in standard order on the worksheet (and ease data entry).

14.7.(b) continued

To analyze the experiment, select **Stat > DOE > Response Surface > Analyze Response Surface Design**. Select "**Terms**" and verify that a full quadratic model (A, B, A^2, B^2, AB) is selected.

Response Surface Regression: Ex14-7y versus x1, x2

The analysis was done using coded units.

Estimated Regression Coefficients for Ex14-7y

Term	Coef	SE Coef	T	P
Constant	13.7273	0.04309	318.580	0.000
x1	0.2980	0.03424	8.703	0.000
x2	-0.4071	0.03424	-11.889	0.000
x1*x1	-0.1249	0.03706	-3.371	0.012
x2*x2	-0.0790	0.03706	-2.132	0.070
x1*x2	0.0550	0.04818	1.142	0.291

...

Analysis of Variance for Ex14-7y

Source	DF	Seq SS	Adj SS	Adj MS	F	P
Regression	5	2.16128	2.16128	0.43226	46.56	0.000
Linear	2	2.01563	2.01563	1.00781	108.54	0.000
Square	2	0.13355	0.13355	0.06678	7.19	0.020
Interaction	1	0.01210	0.01210	0.01210	1.30	0.291
Residual Error	7	0.06499	0.06499	0.00928		
Lack-of-Fit	3	0.03271	0.03271	0.01090	1.35	0.377
Pure Error	4	0.03228	0.03228	0.00807		
Total	12	2.22628				

Estimated Regression Coefficients for Ex14-7y using data in uncoded units

Term	Coef
Constant	13.7273
x1	0.297980
x2	-0.407071
x1*x1	-0.124927
x2*x2	-0.0790085
x1*x2	0.0550000

Values of x_1 and x_2 maximizing the Mooney viscosity can be found from visual examination of the contour and surface plots, or using Minitab's Response Optimizer.

14.7.(b) continued

Stat > DOE > Response Surface > Contour/Surface Plots

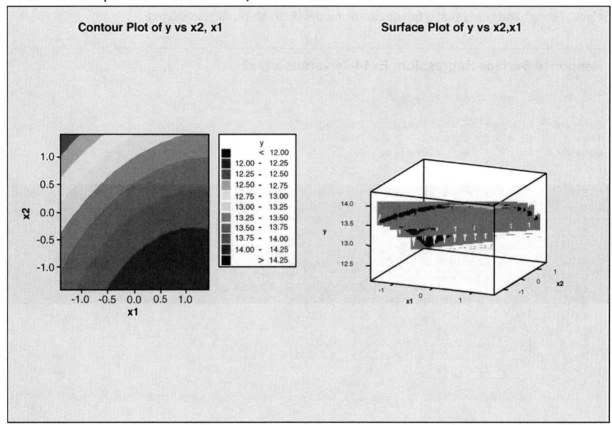

Stat > DOE > Response Surface > Response Optimizer

In Setup, let Goal = maximize, Lower = 10, Target = 20, and Weight = 7.

Optimal		x1	x2
	Hi	1.40	1.40
D	Cur	[1.1857]	[-1.4000]
0.00242	Lo	-1.40	-1.40

y	
Maximum	
y = 14.2287	
d = 0.00242	

From the plots and the optimizer, setting x_1 in a range from 0 to +1.4 and setting x_2 between -1 and -1.4 will maximize viscosity.

14.9.

An article by J. J. Pignatiello, Jr. and J. S. Ramberg in the *Journal of Quality Technology* (Vol. 17, 1985, pp. 198-206) describes the use of a replicated fractional factorial to investigate the effect of five factors on the free height of leaf springs used in an automotive application. The factors are A = furnace temperature, B = heating time, C = transfer time, D = hold down time, and E = quench oil temperature. The data are shown in Table 14E.4.

▪ TABLE 14E.4
The Spring Experiment for Exercise 14.9

A	B	C	D	E			
−	−	−	−	−	7.78,	7.78,	7.81
+	−	−	+	−	8.15,	8.18,	7.88
−	+	−	+	−	7.50,	7.56,	7.50
+	+	−	−	−	7.59,	7.56,	7.75
−	−	+	+	−	7.54,	8.00,	7.88
+	−	+	−	−	7.69,	8.09,	8.06
−	+	+	−	−	7.56,	7.52,	7.44
+	+	+	+	−	7.56,	7.81,	7.69
−	−	−	−	+	7.50,	7.25,	7.12
+	−	−	+	+	7.88,	7.88,	7.44
−	+	−	+	+	7.50,	7.56,	7.50
+	+	−	−	+	7.63,	7.75,	7.56
−	−	+	+	+	7.32,	7.44,	7.44
+	−	+	−	+	7.56,	7.69,	7.62
−	+	+	−	+	7.18,	7.18,	7.25
+	+	+	+	+	7.81,	7.50,	7.59

The design and data are in the Minitab worksheet **Ex14-9.MTW**.

Enter the factor levels and response data into a Minitab worksheet, and then define the experiment using **Stat**
> DOE > Factorial > Define Custom Factorial Design.

(a) Write out the alias structure for this design. What is the resolution of the design?

The defining relation for this half-fraction design is $I = ABCD$ (from examination of the plus and minus signs).

A+BCD	AB+CD	CE+ABDE
B+ACD	AC+BD	DE+ABCE
C+ABD	AD+BC	ABE+CDE
D+ABC	AE+BCDE	ACE+BDE
E	BE+ACDE	ADE+BCE

This is a resolution IV design. All main effects are clear of 2-factor interactions, but some 2-factor interactions are aliased with each other.

14.9. continued

(b) Analyze the data. What factors influence mean free height?

The full model for mean:

Factorial Fit: FreeHeight versus A, B, C, D, E

Estimated Effects and Coefficients for FreeHeight (coded units)

Term	Effect	Coef	SE Coef	T	P	
Constant		7.6256	0.02021	377.41	0.000	
A	0.2421	0.1210	0.02021	5.99	0.000	*
B	-0.1638	-0.0819	0.02021	-4.05	0.000	*
C	-0.0496	-0.0248	0.02021	-1.23	0.229	
D	-0.2387	-0.1194	0.02021	-5.91	0.000	*
E	0.0396	0.0198	0.02021	0.98	0.335	
A*B	-0.0296	-0.0148	0.02021	-0.73	0.469	
A*C	0.0013	0.0006	0.02021	0.03	0.976	
A*D	0.0637	0.0319	0.02021	1.58	0.124	
A*E	-0.0596	-0.0298	0.02021	-1.47	0.150	
B*C	-0.0229	-0.0115	0.02021	-0.57	0.575	
B*D	0.1529	0.0765	0.02021	3.78	0.001	*
B*E	0.0196	0.0098	0.02021	0.48	0.631	
C*D	-0.0329	-0.0165	0.02021	-0.81	0.421	
C*E	0.0021	0.0010	0.02021	0.05	0.959	
D*E	0.0912	0.0456	0.02021	2.26	0.031	*

. . .

Analysis of Variance for FreeHeight (coded units)

Source	DF	Seq SS	Adj SS	Adj MS	F	P
Main Effects	5	1.75734	1.75734	0.351469	17.94	0.000
2-Way Interactions	10	0.50637	0.50637	0.050637	2.58	0.020
Residual Error	32	0.62707	0.62707	0.019596		
Pure Error	32	0.62707	0.62707	0.019596		
Total	47	2.89078				

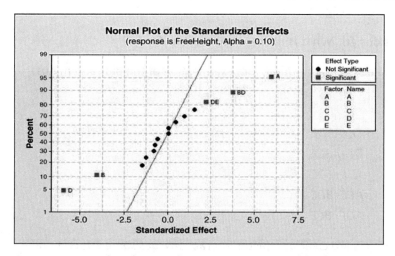

Factors *A*, *B*, and *D* along with interactions *BD* and *DE* (plus *E* to maintain hierarchy) influence mean free height.

14.9. continued

(c) Calculate the range and standard deviation of free height for each run. Is there any indication that any of these factors affects variability in free height?

For Range:

Factorial Fit: Ex14-9Range versus A, B, C, D, E

Estimated Effects and Coefficients for Ex14-9Range (coded units)

Term	Effect	Coef
Constant		0.21937
A	0.11375	0.05688
B	-0.12625	-0.06312
C	0.02625	0.01313
D	-0.01375	-0.00687
E	-0.02125	-0.01063
A*B	0.04375	0.02188
A*C	-0.03375	-0.01688
A*D	-0.00375	-0.00188
A*E	0.13875	0.06938
B*C	0.03625	0.01812
B*D	0.01625	0.00812
B*E	0.04875	0.02438
C*D	-0.13625	-0.06812
C*E	0.03125	0.01562
D*E	0.06125	0.03062

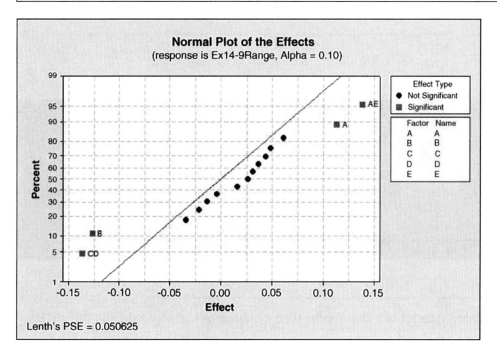

14.9.(c) continued

For Standard Deviation:

Factorial Fit: Ex14-9StdDev versus A, B, C, D, E

```
Estimated Effects and Coefficients for Ex14-9StdDev (coded units)

Term          Effect        Coef
Constant                 0.11744
A            0.06259     0.03129
B           -0.07149    -0.03574
C            0.01057     0.00528
D           -0.00684    -0.00342
E           -0.00468    -0.00234
A*B          0.01540     0.00770
A*C         -0.02185    -0.01093
A*D         -0.00329    -0.00165
A*E          0.07643     0.03822
B*C          0.01906     0.00953
B*D          0.00877     0.00438
B*E          0.01997     0.00999
C*D         -0.07148    -0.03574
C*E          0.01556     0.00778
D*E          0.03536     0.01768
```

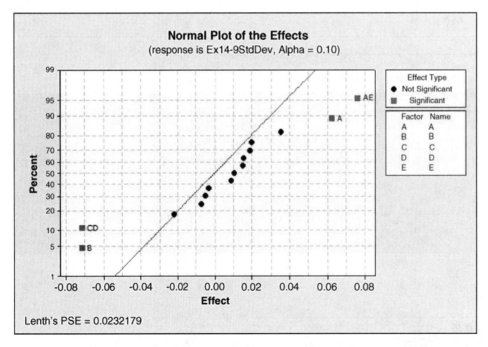

For both models of variability, interactions CE (transfer time × quench oil temperature) and ADE=BCE, along with factors B (heating time) and A (furnace temperature) are significant. Factors C and E are included to keep the models hierarchical.

14.9. continued

(d) Analyze the residuals from this experiment and comment on your findings.

For Mean Height:

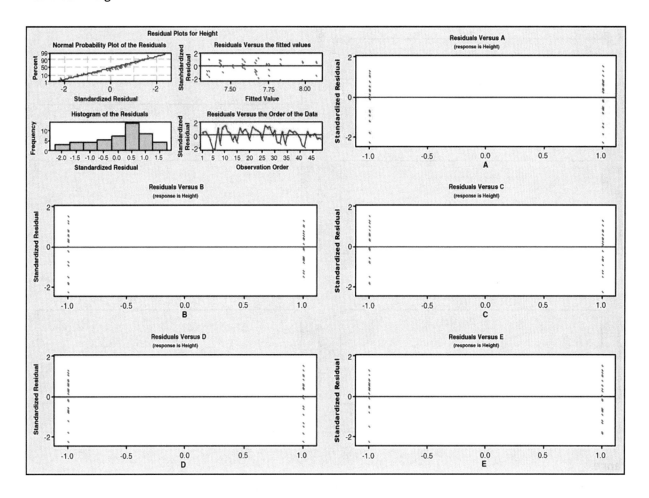

Mean Height
Plot of residuals versus predicted indicates constant variance assumption is reasonable. Normal probability plot of residuals support normality assumption. Plots of residuals versus each factor shows that variance is less at low level of factor E.

14.9.(d) continued

For Range:

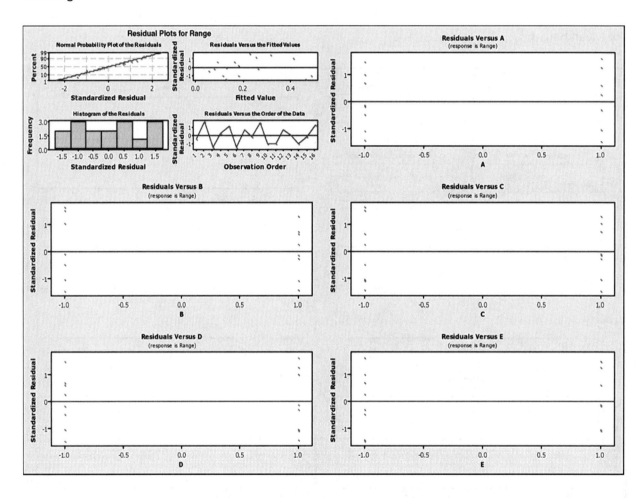

Range

Plot of residuals versus predicted shows that variance is approximately constant over range of predicted values. Residuals normal probability plot indicate normality assumption is reasonable Plots of residuals versus each factor indicate that the variance may be different at different levels of factor D.

14.9.(d) continued

For Standard Deviation:

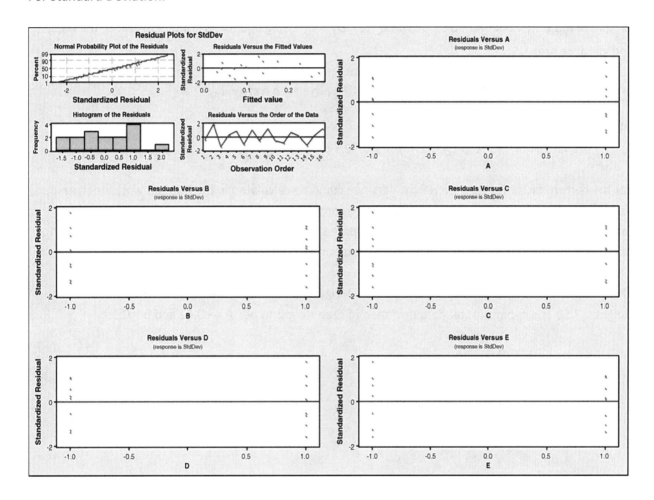

Standard Deviation

Residuals versus predicted plot and residuals normal probability plot support constant variance and normality assumptions. Plots of residuals versus each factor indicate that the variance may be different at different levels of factor D.

(e) This is not the best 16-run design for five factors. A resolution V design can be generated with E = ± ABCD, then none of the 2-factor interactions will be aliased with each other.

14.11.

Consider the leaf spring experiment in Exercise 14.9. Rework this problem, assuming that factors A, B, and C are easy to control but factors D and E are hard to control.

Factors D and E are noise variables. Assume $\sigma_D^2 = \sigma_E^2 = 1$. Using equations (14.6) and (14.7), the mean and variance are:

Mean Free Height = $7.63 + 0.12A - 0.081B$

Variance of Free Height = $\sigma_D^2 (+0.046)^2 + \sigma_E^2 (-0.12 + 0.077B)^2 + \sigma^2$

Using $\hat{\sigma}^2 = MS_E = 0.02$:

Variance of Free Height = $(0.046)^2 + (-0.12 + 0.077B)^2 + 0.02$

For the current factor levels, Free Height Variance could be calculated in the Minitab worksheet, and then contour plots in factors A, B, and D could be constructed using the Graph > Contour Plot functionality. These contour plots could be compared with a contour plot of Mean Free Height, and optimal settings could be identified from visual examination of both plots.

The overlaid contour plot below (constructed in Design-Expert) shows one solution with mean Free Height $\cong 7.50$ and minimum standard deviation of Free Height to be: $A = -0.42$ and $B = 0.99$.

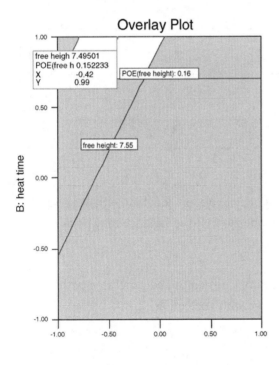

14.13.

The data in Table 14E.6 were collected by a chemical engineer. The response y is filtration time, x_1 is temperature and x_2 is pressure. Fit a second-order model.

■ **TABLE 14E.6**
Chemical Process Experiment
for Exercise 14.13

x_1	x_2	y
-1	-1	54
-1	1	45
1	-1	32
1	1	47
-1.414	0	50
1.414	0	53
0	-1.414	47
0	1.414	51
0	0	41
0	0	39
0	0	44
0	0	42
0	0	40

Since the runs are listed in a patterned (but not standard) order, one approach to solving this exercise is to create a 2-factor central composite design in Minitab with $\alpha = 1.414$, and then enter the data. The design and data are in the Minitab worksheet **Ex14-13.MTW**.

Response Surface Regression: y versus x1, x2

The analysis was done using coded units.
Estimated Regression Coefficients for y

Term	Coef	SE Coef	T	P
Constant	41.200	2.100	19.616	0.000
x1	-1.970	1.660	-1.186	0.274
x2	1.457	1.660	0.878	0.409
x1*x1	3.712	1.781	2.085	0.076
x2*x2	2.463	1.781	1.383	0.209
x1*x2	6.000	2.348	2.555	0.038

...

Analysis of Variance for y

Source	DF	Seq SS	Adj SS	Adj MS	F	P
Regression	5	315.60	315.60	63.119	2.86	0.102
Linear	2	48.02	48.02	24.011	1.09	0.388
Square	2	123.58	123.58	61.788	2.80	0.128
Interaction	1	144.00	144.00	144.000	6.53	0.038
Residual Error	7	154.40	154.40	22.058		
Lack-of-Fit	3	139.60	139.60	46.534	12.58	0.017
Pure Error	4	14.80	14.80	3.700		
Total	12	470.00				

14.13. continued

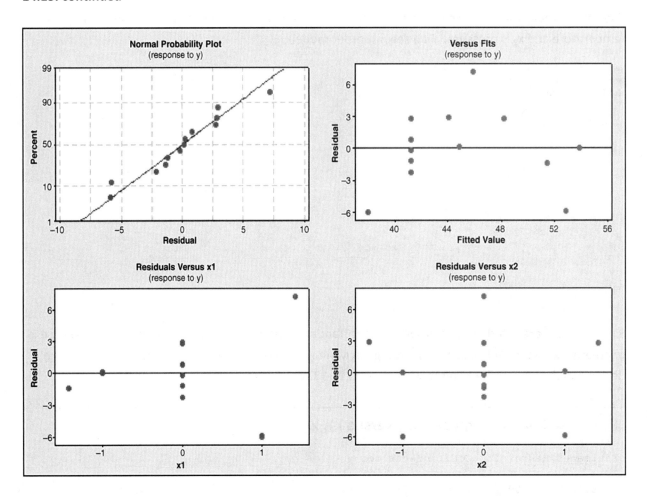

Visual examination of the residuals plots does not indicate any concerns with model assumptions.

14.13. continued

(a) What operating conditions would you recommend if the objective is to minimize the filtration time?

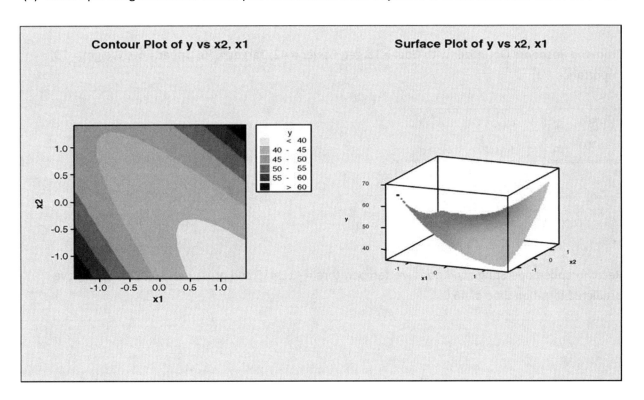

Visual examination of the contour and surface plots indicates that minimum filtration time is achieved near $x_1 \approx = +1.4$ and $x_2 \approx -1.4$.

From the Response Optimizer, with Goal = Minimize, Target = 0, Upper = 55, Weight = 1, Importance = 1:

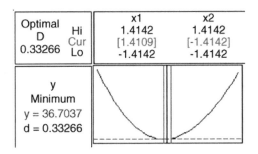

Recommended operating conditions are temperature = +1.4109 and pressure = -1.4142, to achieve predicted filtration time of 36.7.

14.13. continued

(b) What operating conditions would you recommend if the objective is to operate the process at a mean filtration rate very close to 46?

From the Response Optimizer with Goal = Target, Lower = 42, Target = 46, Upper = 50, Weight = 10, Importance = 1:

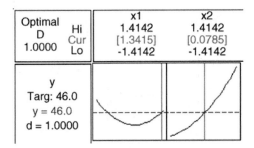

Optimal D 1.0000	Hi Cur Lo	x1 1.4142 [1.3415] -1.4142	x2 1.4142 [0.0785] -1.4142
y Targ: 46.0 y = 46.0 d = 1.0000			

Recommended operating conditions are temperature = +1.3415 and pressure = -0.0785, to achieve predicted filtration time of 46.0.

Chapter 15

Lot-by-Lot Acceptance Sampling for Attributes

LEARNING OBJECTIVES

After completing this chapter you should be able to:

1. Understand the role of acceptance sampling in modern quality control systems
2. Understand the advantages and disadvantages of sampling
3. Understand the difference between attributes and variables sampling plans, and the major types of acceptance-sampling procedures
4. Know how single-, double-, and sequential-sampling plans are used
5. Understand the importance of random sampling
6. Know how to determine the OC curve for a single-sampling plan for attributes
7. Understand the effects of the sampling plan parameters on sampling plan performance
8. Know how to design single-sampling, double-sampling, and sequential sampling plans for attributes
9. Know how rectifying inspection is used
10. Understand the structure and use of MIL STD 105E and its civilian counterpart plans
11. Understand the structure and use of the Dodge-Romig system of sampling plans

IMPORTANT TERMS AND CONCEPTS

100% inspection

Acceptable quality level (AQL)

Acceptance-sampling plan

ANSI/ASQC Z1.4, ISO 2859

AOQL plans

Attributes data

Average outgoing quality

Average outgoing quality limit

Average sample number curve

Average total inspection

Dodge-Romig sampling plans

Double-sampling plan

Ideal OC curve

Lot disposition actions

Lot sentencing

Lot tolerance percent defective (LTPD)

LTPD plans

MIL STD 105E

Multiple-sampling plan

Normal, tightened, and reduced inspection

Operating-characteristic (OC) curve

Random sampling

Rectifying inspection

Sample size code letters

Sequential-sampling plan

Single-sampling plan

Switching rules in MIL STD 105E

Type-A and Type-B OC Curves

Variables data

EXERCISES

Note: Many of the exercises in this chapter are easily solved with spreadsheet application software. The BINOMDIST, HYPGEOMDIST, and graphing functions in Microsoft® Excel® were used for these solutions. Solutions are in the Excel workbook **Chap15.xls**.

15.5.
Draw the type-B OC curve for the single-sampling plan $n = 50$, $c = 1$.

p	f(d=0)	f(d=1)	Pr{d<=c}
0.001	0.951	0.048	0.999
0.002	0.905	0.091	0.995
0.003	0.861	0.129	0.990
0.004	0.818	0.164	0.983
0.005	0.778	0.196	0.974
0.006	0.740	0.223	0.964
0.007	0.704	0.248	0.952
0.008	0.669	0.270	0.939
0.009	0.636	0.289	0.925
0.010	0.605	0.306	0.911
0.011	0.575	0.320	0.895
0.012	0.547	0.332	0.879
0.013	0.520	0.342	0.862
0.014	0.494	0.351	0.845
0.015	0.470	0.358	0.827
0.020	0.364	0.372	0.736
0.025	0.282	0.362	0.644
0.030	0.218	0.337	0.555
0.035	0.168	0.305	0.474
0.040	0.130	0.271	0.400
0.041	0.123	0.264	0.387
0.042	0.117	0.257	0.374
0.043	0.111	0.250	0.361
0.044	0.105	0.243	0.348
0.045	0.100	0.236	0.336
0.050	0.077	0.202	0.279
0.060	0.045	0.145	0.190
0.070	0.027	0.100	0.126
0.080	0.015	0.067	0.083
0.090	0.009	0.044	0.053
0.100	0.005	0.029	0.034
0.110	0.003	0.018	0.021
0.120	0.002	0.011	0.013
0.130	0.001	0.007	0.008
0.140	0.001	0.004	0.005
0.150	0.000	0.003	0.003
0.200	0.000	0.000	0.000
0.250	0.000	0.000	0.000

15.5. continued

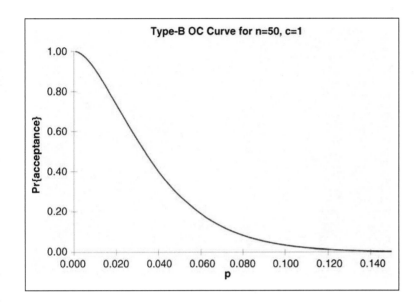

15.7.

Suppose that a product is shipped in lots of size N = 5,000. The receiving inspection procedure used is single sampling with n = 50 and c = 1.

(a) Draw the type-A OC curve for the plan.

N =	5000					
n =	50					
c =	1					
d =	0					
			hypergeometric - Type A			
p		*N*p*	*D*	*f(d=0)*	*f(d=1)*	*Pr{d<=1}*
0.001		5	5	0.951	0.048	0.999
0.002		10	10	0.904	0.092	0.996
0.003		15	15	0.860	0.131	0.991
0.004		20	20	0.818	0.166	0.983
0.005		25	25	0.777	0.197	0.975
0.006		30	30	0.739	0.225	0.964
0.007		35	35	0.703	0.250	0.953
0.008		37.5	37	0.689	0.259	0.948
0.008		40	40	0.668	0.272	0.940
0.009		45	45	0.635	0.291	0.926
0.010		50	50	0.604	0.308	0.911
0.011		55	55	0.574	0.322	0.896
0.012		60	60	0.545	0.334	0.880
0.013		65	65	0.518	0.345	0.863
0.014		70	70	0.492	0.353	0.845
0.015		75	75	0.468	0.360	0.828
0.020		100	100	0.362	0.373	0.736
0.025		125	125	0.280	0.363	0.643
0.030		150	150	0.216	0.338	0.554
0.035		175	175	0.167	0.306	0.473
0.040		200	200	0.129	0.271	0.399
0.045		225	225	0.099	0.235	0.334
0.050		250	250	0.076	0.202	0.278
0.060		300	300	0.045	0.144	0.189
0.070		350	350	0.026	0.099	0.125
0.075		375	375	0.020	0.081	0.101
0.080		400	400	0.015	0.067	0.082
0.090		450	450	0.009	0.044	0.052
0.100		500	500	0.005	0.028	0.033
0.110		550	550	0.003	0.018	0.021
0.120		600	600	0.002	0.011	0.013
0.130		650	650	0.001	0.007	0.008
0.140		700	700	0.001	0.004	0.005
0.150		750	750	0.000	0.003	0.003
0.200		1000	1000	0.000	0.000	0.000
0.250		1250	1250	0.000	0.000	0.000

15.7.(a) continued

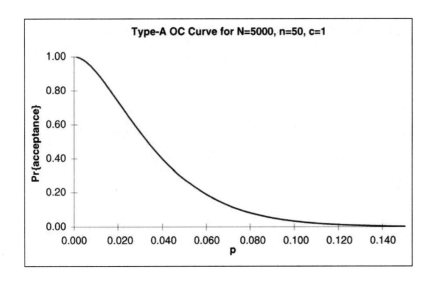

$P_a\ (d = 35) = 0.9521$, or $\alpha \cong 0.05$

$P_a\ (d = 375) = 0.10133$, or $\beta \cong 0.10$

15.7. continued

(b) Draw the type-B OC curve for this plan and compare it to the type-A OC curve found in part (a).

N =	5000			
n =	50			
c =	1			
d =	0	1		
			binomial - Type B	
p			**Pr{d<=1}**	
0.001			0.999	
0.002			0.995	
0.003			0.990	
0.004			0.983	
0.005			0.974	
0.006			0.964	
0.007			0.952	
0.008			0.946	
0.008			0.939	
0.009			0.925	
0.010			0.911	
0.011			0.895	
0.012			0.879	
0.013			0.862	
0.014			0.845	
0.015			0.827	
0.020			0.736	
0.025			0.644	
0.030			0.555	
0.035			0.474	
0.040			0.400	
0.045			0.336	
0.050			0.279	
0.060			0.190	
0.070			0.126	
0.075			0.103	
0.080			0.083	
0.090			0.053	
0.100			0.034	
0.110			0.021	
0.120			0.013	
0.130			0.008	
0.140			0.005	
0.150			0.003	
0.200			0.000	
0.250			0.000	

15.7.(b) continued

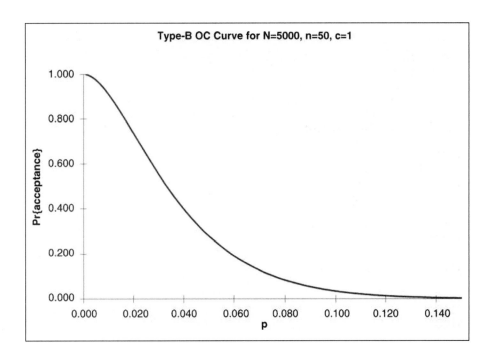

P_a (p = 0.007) = 0.9521, or $\alpha \cong 0.05$

P_a (p = 0.075) = 0.10133, or $\beta \cong 0.10$

(c) Which curve is appropriate for this situation?

Based on values for α and β, the difference between the two curves is small; either is appropriate.

15.9.

Find a single-sampling plan for which p_1 = 0.05, α = 0.05, p_2 = 0.15, and β = 0.10.

$p_1 = 0.05; 1-\alpha = 1-0.05 = 0.95; p_2 = 0.15; \beta = 0.10$

From the binomial nomograph, the sampling plan is n = 80 and c = 7.

15.11.

A company uses the following acceptance-sampling procedure. A sample equal to 10% of the lot is taken. If 2% or less of the items in the sample are defective, the lot is accepted; otherwise, it is rejected. If submitted lots vary in size from 5,000 to 10,000 units, what can you say about the protection by this plan? If 0.05 is the desired LTPD, does this scheme offer reasonable protection to the consumer?

15.11. continued

LTPD =		0.05				
	N1 =	5000		**N2 =**	10000	
	n1 =	500		**n1 =**	1000	
	pmax =	0.0200		**pmax =**	0.0200	
	cmax =	10		**cmax =**	20	
	binomial			**binomial**		
p	**Pr{d<=10}**	**Pr{reject}**		**Pr{d<=20}**	**Pr{reject}**	
0.001	1.000	0.000		1.000	0.000	
0.002	1.000	0.000		1.000	0.000	
0.003	1.000	0.000		1.000	0.000	
0.004	1.000	0.000		1.000	0.000	
0.005	1.000	0.000		1.000	0.000	
0.006	1.000	0.000		1.000	0.000	
0.007	0.999	0.001		1.000	0.000	
0.008	0.998	0.002		1.000	0.000	
0.008	0.997	0.003		1.000	0.000	
0.009	0.994	0.006		1.000	0.000	
0.010	0.987	0.013		0.999	0.001	
0.011	0.975	0.025		0.996	0.004	
0.012	0.958	0.042		0.989	0.011	
0.013	0.934	0.066		0.976	0.024	
0.014	0.903	0.097		0.953	0.047	
0.015	0.864	0.136		0.919	0.081	
0.020	0.583	0.417		0.559	0.441	
0.025	0.294	0.706		0.182	0.818	
0.030	0.115	0.885		0.033	0.967	
0.035	0.036	0.964		0.004	0.996	
0.040	0.010	0.990		0.000	1.000	
0.045	0.002	0.998		0.000	1.000	
0.050	0.000	1.000		0.000	1.000	
0.060	0.000	1.000		0.000	1.000	
0.070	0.000	1.000		0.000	1.000	
0.075	0.000	1.000		0.000	1.000	
0.080	0.000	1.000		0.000	1.000	
0.090	0.000	1.000		0.000	1.000	
0.100	0.000	1.000		0.000	1.000	
0.110	0.000	1.000		0.000	1.000	
0.120	0.000	1.000		0.000	1.000	
0.130	0.000	1.000		0.000	1.000	
0.140	0.000	1.000		0.000	1.000	
0.150	0.000	1.000		0.000	1.000	
0.200	0.000	1.000		0.000	1.000	
0.250	0.000	1.000		0.000	1.000	

Different sample sizes offer different levels of protection. For $N = 5,000$, $P_a(p = 0.025) = 0.294$; while for $N = 10,000$, $P_a(p = 0.025) = 0.182$. Also, the consumer is protected from a LTPD = 0.05 by $P_a(N = 5,000) = 0.00046$ and $P_a(N = 10,000) = 0.00000$, but pays for the high probability of rejecting acceptable lots like those with $p = 0.025$.

15.13.

Consider the single-sampling plan found in Exercise 15.8. Suppose that lots of $N = 2,000$ are submitted. Draw the ATI curve for this plan. Draw the AOQ curve and find the AOQL.

$n = 35$; $c = 1$; $N = 2,000$

$$ATI = n + (1 - P_a)(N - n)$$
$$= 35 + (1 - P_a)(2000 - 35)$$
$$= 2000 - 1965 P_a$$

$$AOQ = \frac{P_a p(N - n)}{N}$$
$$= (1965/2000) P_a p$$

$$AOQL = 0.0234$$

15.13. continued

p	binomial Pa=Pr{d<=1}	ATI	AOQ	
N =	2000			
n =	35			
c =	1			
0.001	0.999	36	0.0010	
0.002	0.998	39	0.0020	
0.003	0.995	45	0.0029	
0.004	0.991	52	0.0039	
0.005	0.987	61	0.0048	
0.006	0.981	72	0.0058	
0.007	0.975	84	0.0067	
0.008	0.972	91	0.0072	
0.008	0.968	98	0.0076	
0.009	0.960	113	0.0085	
0.010	0.952	129	0.0094	
0.011	0.943	146	0.0102	
0.012	0.934	165	0.0110	
0.013	0.924	184	0.0118	
0.014	0.914	204	0.0126	
0.015	0.903	225	0.0133	
0.020	0.845	339	0.0166	
0.025	0.782	463	0.0192	
0.030	0.717	591	0.0211	
0.035	0.652	718	0.0224	
0.040	0.589	843	0.0231	
0.045	0.529	961	0.0234	AOQL
0.050	0.472	1072	0.0232	
0.060	0.371	1271	0.0219	
0.070	0.287	1437	0.0197	
0.075	0.251	1507	0.0185	
0.080	0.218	1571	0.0172	
0.090	0.164	1677	0.0145	
0.100	0.122	1760	0.0120	
0.110	0.090	1823	0.0097	
0.120	0.066	1871	0.0078	
0.130	0.048	1906	0.0061	
0.140	0.034	1933	0.0047	
0.150	0.024	1952	0.0036	
0.200	0.004	1992	0.0008	
0.250	0.001	1999	0.0001	
0.300	0.000	2000	0.0000	

15.13. continued

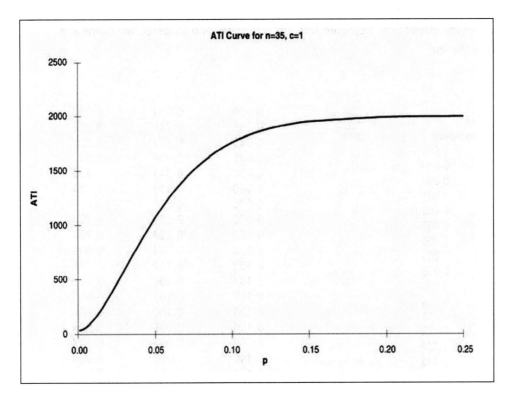

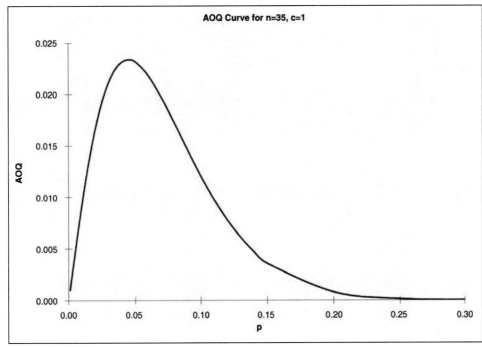

15.15.

Suppose that a supplier ships components in lots of size 5,000. A single-sampling plan with $n = 50$ and $c = 2$ is being used for receiving inspection. Rejected lots are screened, and all defective items are reworked and returned to the lot.

N = 5000	n =	50
	c =	2
	binomial	
p	Pa=Pr{d<=2}	Pr{reject}
0.001	1.000	0.000
0.002	1.000	0.000
0.003	1.000	0.000
0.004	0.999	0.001
0.005	0.998	0.002
0.006	0.997	0.003
0.007	0.995	0.005
0.008	0.994	0.006
0.008	0.992	0.008
0.009	0.990	0.010
0.010	0.986	0.014
0.011	0.982	0.018
0.012	0.978	0.022
0.013	0.973	0.027
0.014	0.967	0.033
0.015	0.961	0.039
0.016	0.954	0.046
0.017	0.947	0.053
0.018	0.939	0.061
0.019	0.930	0.070
0.020	0.922	0.078
0.021	0.912	0.088
0.022	0.902	0.098
0.023	0.892	0.108
0.024	0.882	0.118
0.025	0.871	0.129

p	Pa	Pr{reject}
0.030	0.811	0.189
0.035	0.745	0.255
0.040	0.677	0.323
0.045	0.608	0.392
0.050	0.541	0.459
0.060	0.416	0.584
0.070	0.311	0.689
0.075	0.266	0.734
0.080	0.226	0.774
0.090	0.161	0.839
0.100	0.112	0.888
0.101	0.108	0.892
0.102	0.104	0.896
0.103	0.100	0.900
0.104	0.096	0.904
0.105	0.093	0.907
0.110	0.076	0.924
0.120	0.051	0.949
0.130	0.034	0.966
0.140	0.022	0.978
0.150	0.014	0.986
0.200	0.001	0.999
0.250	0.000	1.000
0.300	0.000	1.000
0.350	0.000	1.000
0.400	0.000	1.000
0.450	0.000	1.000
0.500	0.000	1.000

15.15. continued

(a) Draw the OC curve for this plan.

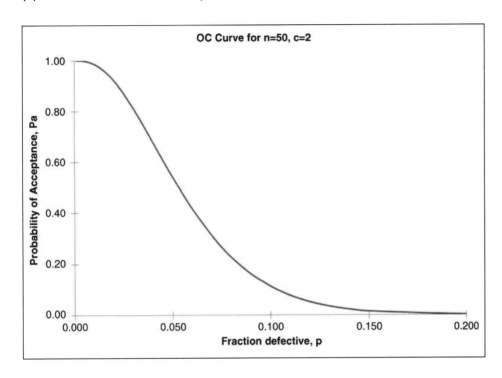

(b) Find the level of lot quality that will be rejected 90% of the time.

$p = 0.1030$ will be rejected about 90% of the time.

(c) Management has objected to the use of the above sampling procedure and wants to use a plan with an acceptance number $c = 0$, arguing that this is more consistent with their zero-defects program. What do you think of this?

A zero-defects sampling plan, with acceptance number $c = 0$, will be extremely hard on the vendor because the P_a is low even if the lot fraction defective is low. Generally, quality improvement begins with the manufacturing process control, not the sampling plan.

15.15. continued

(d) Design a single-sampling plan with $c = 0$ that will give a 0.90 probability of rejection of lots having the quality level found in part (b). Note that the two plans are now matched at the LTPD point. Draw the OC curve for this plan and compare it to the one for $n = 50$, $c = 2$ in part (a).

From the nomograph, select $n = 20$, yielding $P_a = 1 - 0.11372 = 0.88638 \approx 0.90$. The OC curve for this zero-defects plan is much steeper.

N = 5000	n =	20
	c =	0
	binomial	
p	Pa=Pr{d<=0}	Pr{reject}
0.001	0.980	0.020
0.002	0.961	0.039
0.003	0.942	0.058
0.004	0.923	0.077
0.005	0.905	0.095
0.006	0.887	0.113
0.007	0.869	0.131
0.008	0.860	0.140
0.008	0.852	0.148
0.009	0.835	0.165
0.010	0.818	0.182
0.011	0.802	0.198
0.012	0.785	0.215
0.013	0.770	0.230
0.014	0.754	0.246
0.015	0.739	0.261
0.016	0.724	0.276
0.017	0.710	0.290
0.018	0.695	0.305
0.019	0.681	0.319
0.020	0.668	0.332
0.021	0.654	0.346
0.022	0.641	0.359
0.023	0.628	0.372
0.024	0.615	0.385
0.025	0.603	0.397

p	Pa	Pr{reject}
0.030	0.544	0.456
0.035	0.490	0.510
0.040	0.442	0.558
0.045	0.398	0.602
0.050	0.358	0.642
0.060	0.290	0.710
0.070	0.234	0.766
0.075	0.210	0.790
0.080	0.189	0.811
0.090	0.152	0.848
0.100	0.122	0.878
0.101	0.119	0.881
0.102	0.116	0.884
0.103	0.114	0.886
0.104	0.111	0.889
0.105	0.109	0.891
0.110	0.097	0.903
0.120	0.078	0.922
0.130	0.062	0.938
0.140	0.049	0.951
0.150	0.039	0.961
0.200	0.012	0.988
0.250	0.003	0.997
0.300	0.001	0.999
0.350	0.000	1.000
0.400	0.000	1.000
0.450	0.000	1.000
0.500	0.000	1.000

15.15.(d) continued

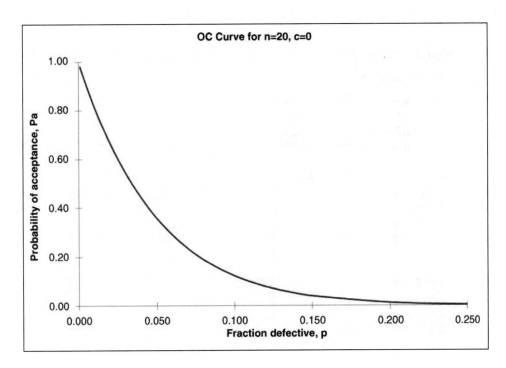

(e) Suppose that incoming lots are 0.5% nonconforming. What is the probability of rejecting these lots under both plans? Calculate the ATI at this point for both plans. Which plan do you prefer? Why?

$\text{Pr}\{\text{reject} \mid p = 0.005, c = 0\} = 0.09539$

$\text{Pr}\{\text{reject} \mid p = 0.005, c = 2\} = 0.00206$

$\text{ATI}_{c=0} = n + (1 - P_a)(N - n) = 20 + (0.09539)(5000 - 20) = 495$

$\text{ATI}_{c=2} = 50 + (0.00206)(5000 - 50) = 60$

The $c = 2$ plan is preferred because the $c = 0$ plan will reject good lots 10% of the time.

15.17.

(a) Derive an item-by-item sequential-sampling plan for which $p_1 = 0.01$, $\alpha = 0.05$, $p_2 = 0.10$, and $\beta = 0.10$.

$k = 1.0414; h_1 = 0.9389; h_2 = 1.2054; s = 0.0397$

$X_A = -0.9389 + 0.0397n; X_R = 1.2054 + 0.0397n$

p1 =	0.01		k =	1.0414
p2 =	0.10		h1 =	0.9389
alpha =	0.05		h2 =	1.2054
beta =	0.10		s =	0.0397

n	XA	XR	Acc	Rej
1	-0.899	1.245	n/a	2
2	-0.859	1.285	n/a	2
3	-0.820	1.325	n/a	2
4	-0.780	1.364	n/a	2
5	-0.740	1.404	n/a	2
...				
20	-0.144	2.000	n/a	2
21	-0.104	2.040	n/a	3
22	-0.064	2.080	n/a	3
23	-0.025	2.120	n/a	3
24	0.015	2.159	0	3
25	0.055	2.199	0	3
...				
45	0.850	2.994	0	3
46	0.890	3.034	0	4
47	0.929	3.074	0	4
48	0.969	3.113	0	4
49	1.009	3.153	1	4
50	1.049	3.193	1	4

The sampling plan is $n = 49$; Acc = 1; Rej = 4.

(b) Draw the OC curve for this plan.

Three points on the OC curve are:

$p_1 = 0.01; P_a = 1 - \alpha = 0.95$

$p = s = 0.0397; P_a = \dfrac{h_2}{h_1 + h_2} = \dfrac{1.2054}{0.9389 + 1.2054} = 0.5621$

$p_2 = 0.10; P_a = \beta = 0.10$

15.19.

Consider rectifying inspection for single sampling. Develop an AOQ equation assuming that all defective items are removed but not replaced with good ones.

$$AOQ = \left[P_a \times p \times (N-n)\right] / \left[N - P_a \times (np) - (1 - P_a) \times (Np)\right]$$

15.21.

Repeat Exercise 15.20, using general inspection level I. Discuss the differences in the various sampling plans.

$N = 3000$, AQL = 1%
General level I
Normal sampling plan: Sample size code letter = H, $n = 50$, Ac = 1, Re = 2
Tightened sampling plan: Sample size code letter = J, $n = 80$, Ac = 1, Re = 2
Reduced sampling plan: Sample size code letter = H, $n = 20$, Ac = 0, Re = 2

15.23.

MIL STD 105E is being used to inspect incoming lots of size $N = 5,000$. Single sampling, general inspection level II, and an AQL of 0.65% are being used.

(a) Find the normal, tightened, and reduced inspection plans.

$N = 5000$, AQL = 0.65%; General level II; Sample size code letter = L
Normal sampling plan: $n = 200$, Ac = 3, Re = 4
Tightened sampling plan: $n = 200$, Ac = 2, Re = 3
Reduced sampling plan: $n = 80$, Ac = 1, Re = 4

15.23. continued

(b) Draw the OC curves of the normal, tightened, and reduced inspection plans on the same graph.

N = 5000	normal	tightened	reduced
n =	200	200	80
c =	3	2	1
p	Pa=Pr{d<=3}	Pa=Pr{d<=2}	Pa=Pr{d<=1}
0.001	1.000	0.999	0.997
0.002	0.999	0.992	0.989
0.003	0.997	0.977	0.976
0.004	0.991	0.953	0.959
0.005	0.981	0.920	0.939
0.006	0.967	0.880	0.916
0.007	0.947	0.834	0.892
0.008	0.935	0.809	0.879
0.008	0.922	0.784	0.865
0.009	0.892	0.731	0.838
0.010	0.858	0.677	0.809
0.011	0.820	0.622	0.780
0.012	0.779	0.569	0.751
0.013	0.737	0.517	0.721
0.014	0.692	0.468	0.691
0.015	0.647	0.421	0.662
0.016	0.602	0.378	0.633
0.017	0.558	0.337	0.605
0.018	0.514	0.300	0.577
0.019	0.472	0.266	0.549
0.020	0.431	0.235	0.523
0.021	0.393	0.207	0.497
0.022	0.357	0.182	0.472
0.023	0.323	0.159	0.448
0.024	0.291	0.139	0.425
0.025	0.261	0.121	0.403
0.030	0.147	0.059	0.304
0.035	0.078	0.028	0.226
0.040	0.040	0.012	0.165
0.045	0.019	0.005	0.120
0.050	0.009	0.002	0.086
0.060	0.002	0.000	0.043
0.070	0.000	0.000	0.021
0.075	0.000	0.000	0.015
0.080	0.000	0.000	0.010
0.090	0.000	0.000	0.005
0.100	0.000	0.000	0.002

15.23.(b) continued

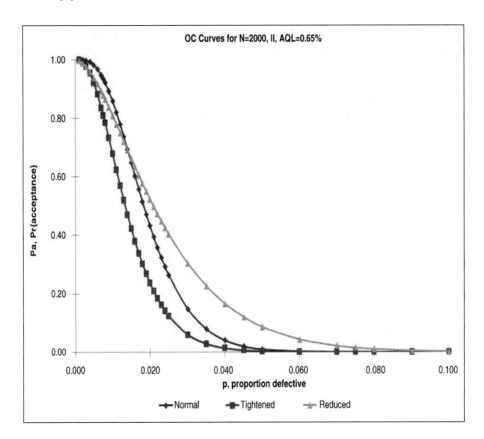

15.25.

We wish to find a single-sampling plan for a situation where lots are shipped from a supplier. The supplier's process operates at a fallout level of 0.50% defective. We want the AOQL from the inspection activity to be 3%.

(a) Find the appropriate Dodge–Romig plan.

Dodge-Romig single sampling, AOQL = 3%, average process fallout = p = 0.50% defective
Minimum sampling plan that meets the quality requirements is $50{,}001 \leq N \leq 100{,}000$; $n = 65$; $c = 3$.

15.25. continued

(b) Draw the OC curve and the ATI curve for this plan. How much inspection will be necessary, on the average, if the supplier's process operates close to the average fallout level?

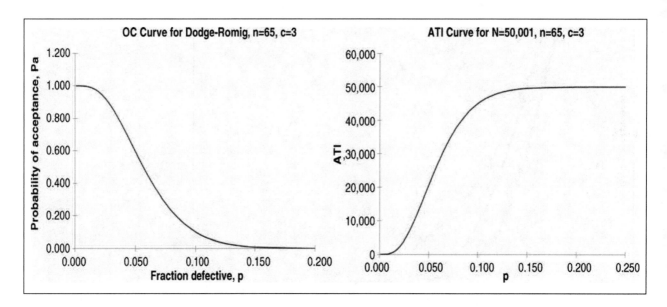

let $N = 50,001$

$P_a = Binom(3,65,0.005) = 0.99967$

$ATI = n + (1 - P_a)(N - n) = 65 + (1 - 0.99967)(50,001 - 65) = 82$

On average, if the vendor's process operates close to process average, the average inspection required will be 82 units.

(c) What is the LTPD protection for this plan?

LTPD = 10.3%

Chapter 16

Other Acceptance-Sampling Techniques

LEARNING OBJECTIVES

After completing this chapter you should be able to:

1. Understand the difference between attributes and variables sampling plans
2. Understand the advantages and disadvantages of variables sampling
3. Understand the two major types of variables sampling plans
4. Know how to design a variables sampling plan with a specified OC
5. Understand the structure and use of MIL STD 414 and its civilian counterpart plans
6. Understand the differences between the MIL STD 414 and ANSI/ASQC Z1.9 sampling plans
7. Understand how chain sampling plans are designed and used
8. Understand how continuous sampling plans are designed and used
9. Understand how skip-lot sampling plans are designed and used

IMPORTANT TERMS AND CONCEPTS

Acceptable quality level (AQL)

ANSI/ASQC Z1.9

Average outgoing quality limit (AOQL)

Average sample number (ASN)

Chain sampling

Clearance number

Continuous sampling plans

MIL STD 414

Normal distribution in variables sampling

Normal, tightened, and reduced inspection

Operating-characteristic (OC) curve

Sample-size code letters

Skip-lot sampling plans

Switching between normal, tightened, and reduced inspection

Variables data

Zero-acceptance-number plans

EXERCISES

Note:

- Many of the exercises in this chapter are easily solved with spreadsheet application software. The BINOMDIST and graphing functions in Microsoft® Excel® were used for these solutions. Solutions are in the Excel workbook **Chap 16.xls.**
- Some exercises require a full copy of MIL-STD-414. The document is available on the Montgomery SQC website: www.wiley.com/college/montgomery, and also at: http://archive.org/details/MIL-STD-414

16.1.

The density of a plastic part used in a cellular telephone is required to be at least 0.70 g/cm3. The parts are supplied in large lots, and a variables sampling plan is to be used to sentence the lots. It is desired to have $p_1 = 0.02$, $p_2 = 0.10$, $\alpha = 0.10$, and $\beta = 0.05$. The variability of the manufacturing process is unknown but will be estimated by the sample standard deviation.

LSL = 0.70 g/cm^3, $p_1 = 0.02$; $1 - \alpha = 1 - 0.10 = 0.90$; $p_2 = 0.10$; $\beta = 0.05$

(a) Find an appropriate variables sampling plan, using Procedure 1.

From the variables nomograph (Figure 16-2), the sampling plan is $n = 35$; $k = 1.7$. Calculate $\bar{x}$ and S. Accept the lot if $\left[Z_{LSL} = (\bar{x} - LSL)/S \right] \geq 1.7$.

(b) Suppose that a sample of the appropriate size was taken, and $\bar{x} = 0.73$ and $s = 1.05 \times 10^{-2}$. Should the lot be accepted or rejected?

$\bar{x} = 0.73; S = 1.05 \times 10^{-2}$

$\left[Z_{LSL} = (0.73 - 0.70)/(1.05 \times 10^{-2}) = 2.8571 \right] \geq 1.7$

Accept the lot.

16.1. continued

(c) Sketch the OC curve for this sampling plan. Find the probability of accepting lots that are 5% defective.

$P_a\{p = 0.05\} \approx 0.38$ (from nomograph, Figure 16-2)

From variables nomograph		
n =	35	
k =	1.7	
	p	Pr{accept}
	0.010	0.988
	0.016	0.945
p1	0.020	0.900 1-alpha
	0.025	0.820
	0.030	0.730
	0.040	0.560
	0.050	0.400
	0.070	0.190
p2	0.100	0.050 beta
	0.150	0.005
	0.190	0.001

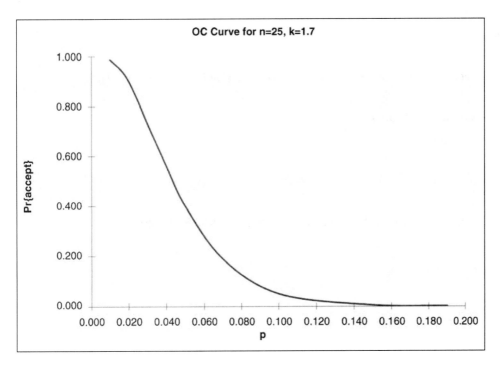

OC Curve for n=25, k=1.7

16.3.

Describe how rectifying inspection can be used with variables sampling. What are the appropriate equations for the AOQ and the ATI, assuming single sampling and requiring that all defective items found in either sampling or 100% inspection are replaced by good ones?

The equations do not change: AOQ $= P_a p (N - n) / N$ and ATI $= n + (1 - P_a) (N - n)$. The design of a variables plan in rectifying inspection is somewhat different from the attribute plan design, and generally involves some trial-and-error search.

For example, for a given AOQL $= P_a p_m (N - n) / N$ (where p_m is the value of p that maximizes AOQ), we know n and k are related, because both P_a and p_m are functions of n and k. Suppose n is arbitrarily specified. Then a k can be found to satisfy the AOQL equation. No convenient mathematical method exists to do this, and special Romig tables are usually employed. Now, for a specified process average, n and k will define P_a. Finally, ATI is found from the above equation. Repeat until the n and k that minimize ATI are found.

16.5.

How does the sample size found in Exercise 16.4 compare with what would have been used under MIL STD 105E?

Under MIL STD 105E, Inspection level II, Sample size code letter = L (Tables 15-4, -5, -6, -7):

	Normal	Tightened	Reduced
n	200	200	80
Ac	7	5	3
Re	8	6	6

The MIL STD 414 sample sizes are considerably smaller than those for MIL STD 105E.

16.7.

A soft-drink bottler purchases nonreturnable glass bottles from a supplier. The lower specification on bursting strength in the bottles is 225 psi. The bottler wishes to use variables sampling to sentence the lots and has decided to use an AQL of 1%. Find an appropriate set of normal and tightened sampling plans from the standard. Suppose that a lot is submitted, and the sample results yield $\bar{x} = 225$ and $s = 10$. Determine the disposition of the lot using Procedure 1. The lot size is $N = 100,000$.

LSL = 225psi, AQL = 1%, N = 100,000
Assume inspection level IV, sample size code letter = O (Table 16-1)
Normal sampling: n = 100, k = 2.00 (Table 16-2)
Tightened sampling: n = 100, k = 2.14 (Table 16-2)
Assume normal sampling is in effect.
$\bar{x} = 255; S = 10$
$\left[Z_{LSL} = (\bar{x} - LSL)/S = (255 - 225)/10 = 3.000 \right] > 2.00$, so accept the lot.

16.9.

A standard of 0.3 ppm has been established for formaldehyde emission levels in wood products. Suppose that the standard deviation of emissions in an individual board is $\sigma = 0.10$ ppm. Any lot that contains 1% of its items above 0.3 ppm is considered acceptable. Any lot that has 8% or more of its items above 0.3 ppm is considered unacceptable. Good lots are to be accepted with probability 0.95, and bad lots are to be rejected with probability 0.90.

target = 3ppm; $\sigma = 0.10$ppm; p_1 = 1% = 0.01; p_2 = 8% = 0.08

(a) Derive a variables-sampling plan for this situation.

$1 - \alpha = 0.95$; $\beta = 1 - 0.90 = 0.10$
From the nomograph (Figure 16-2), the sampling plan is n = 30 and k = 1.8.

(b) Using the 1% nonconformance level as an AQL, and assuming that lots consist of 5,000 panels, find an appropriate set of sampling plans from MIL STD 414, assuming σ is unknown. Compare the sample sizes and the protection that both producer and consumer obtain from this plan with the plan derived in part (a).

Note: Tables from MIL-STD-414 required for this exercise.

AQL = 1%; N = 5000; σ unknown
Double specification limit, assume inspection level IV

16.9.(b) continued

From Table A-2: sample size code letter = M
From Tables B-3 and B-4:

> Normal: $n = 50$, M = 2.49 ($k = 1.93$) (Table B-3)
> Tightened: $n = 50$, M = 1.71 ($k = 2.08$) (Table B-3)
> Reduced: $n = 20$, M = 4.09 ($k = 1.69$) (Table B-4)

σ known allows smaller sample sizes than σ unknown.

(c) Find an attributes sampling plan that has the same OC curve as the variables sampling plan derived in part (a). Compare the sample sizes required for equivalent protection. Under what circumstances would variables sampling be more economically efficient?

$1 - \alpha = 0.95$; $\beta = 0.10$; $p_1 = 0.01$; $p_2 = 0.08$
From binomial nomograph (for attributes, Figure 15-9): $n = 60$, $c = 2$

The sample size is slightly larger than required for the variables plan (a). Variables sampling would be more efficient if σ were known.

(d) Using the 1% nonconforming as an AQL, find an attributes-sampling plan from MIL STD 105E. Compare the sample sizes and the protection obtained from this plan with the plans derived in parts (a), (b), and (c).

AQL = 1%; $N = 5,000$
Assume inspection level II: sample size code letter = L (Table 15-4)
Normal: $n = 200$, Ac = 5, Re = 6 (Table 15-5)
Tightened: $n = 200$, Ac = 3, Re = 4 (Table 15-6)
Reduced: $n = 80$, Ac = 2, Re = 5 (Table 15-7)

The sample sizes required are much larger than for the other plans.

16.11.
An electronics manufacturer buys memory devices in lots of 30,000 from a supplier. The supplier has a long record of good quality performance, with an average fraction defective of approximately 0.10%. The quality engineering department has suggested using a conventional acceptance-sampling plan with $n = 32$, $c = 0$.

$N = 30,000$; average process fallout = 0.10% = 0.001, $n = 32$, $c = 0$

16.11. continued

(a) Draw the OC curve of this sampling plan.

N = 30,000	n =	32
	c =	0
	binomial	

p	Pa	Pr{reject}
0.0010	0.9685	0.0315
0.0015	0.9531	0.0469
0.0020	0.9379	0.0621
0.0030	0.9083	0.0917
0.0040	0.8796	0.1204
0.0050	0.8518	0.1482
0.0060	0.8248	0.1752
0.0070	0.7987	0.2013
0.0080	0.7733	0.2267
0.0090	0.7488	0.2512
0.0100	0.7250	0.2750
0.0150	0.6165	0.3835
0.0200	0.5239	0.4761
0.0250	0.4448	0.5552
0.0300	0.3773	0.6227
0.0350	0.3198	0.6802
0.0400	0.2708	0.7292
0.0450	0.2291	0.7709

p	Pa	Pr{reject}
0.0500	0.1937	0.8063
0.0550	0.1636	0.8364
0.0600	0.1381	0.8619
0.0650	0.1164	0.8836
0.0700	0.0981	0.9019
0.0750	0.0825	0.9175
0.0800	0.0694	0.9306
0.0850	0.0583	0.9417
0.0900	0.0489	0.9511
0.0950	0.0410	0.9590
0.1000	0.0343	0.9657
0.1500	0.0055	0.9945
0.2000	0.0008	0.9992
0.2500	0.0001	0.9999
0.3000	0.0000	1.0000
0.3500	0.0000	1.0000

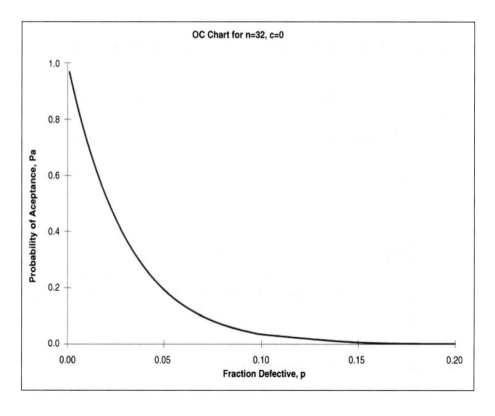

16.11. continued

(b) If lots are of a quality that is near the supplier's long-term process average, what is the average total inspection at that level of quality?

$$ATI = n + (1 - P_a)(N - n)$$
$$= 32 + (1 - 0.9685)(30000 - 32)$$
$$= 976$$

(c) Consider a chain-sampling plan with $n = 32$, $c = 0$, and $i = 3$. Contrast the performance of this plan with the conventional sampling plan with $n = 32$, $c = 0$.

Chain-sampling: $n = 32$, $c = 0$, $i = 3$, $p = 0.001$

$$P_a = P(0,n) + P(1,n)[P(0,n)]^i$$
$$P(0,n) = P(0,32) = 0.9685$$
$$P(1,n) = P(1,32) = 0.0310$$
$$P_a = 0.9685 + (0.0310)(0.9685)^3 = 0.9967$$

$$ATI = 32 + (1 - 0.9967)(30000 - 32) = 131$$

Compared to conventional sampling, the P_a for chain sampling is slightly larger, but the average number inspected is much smaller.

(d) How would the performance of this chain-sampling plan change if we substituted $i = 4$ in part (c)?

$P_a = 0.9958$, there is little change in performance by increasing i.

$$ATI = 32 + (1 - 0.9958)(30000 - 32) = 158$$

16.13.
A chain-sampling plan is used for the inspection of lots of size $N = 500$. The sample size is $n = 6$. If the sample contains no defectives, the lot is accepted. If one defective is found, the lot is accepted provided that the samples from the four previous lots are free of defectives. Determine the probability of acceptance of a lot that is 2% defective.

$N = 500$, $n = 6$
If $c = 0$, accept. If $c = 1$, accept if $i = 4$. Need to find $P_a\{p = 0.02\}$
$$P_a = P(0,6) + P(1,6)[P(0,6)]^4 = 0.88584 + 0.10847(0.88584)^4 = 0.95264$$

16.15.

For the sampling plans developed in Exercise 16.14, compare the plans' performance in terms of average fraction inspected, given that the process is in control at an average fallout level of 0.15%. Compare the plans in terms of their operating-characteristic curves.

Average process fallout, $p = 0.15\% = 0.0015$ and $q = 1 - p = 0.9985$

1. $f = \frac{1}{2}$ and $i = 140$: $u = 156$, $v = 1333$, AFI = 0.5523, $P_a = 0.8953$
2. $f = 1/10$ and $i = 550$: $u = 856$, $v = 6667$, AFI = 0.2024, $P_a = 0.8863$
3. $f = 1/100$ and $i = 1302$: $u = 4040$, $v = 66,667$, AFI = 0.0666, $P_a = 0.9429$

AOQL=0.198%		$p =$	0.0015 $q =$	0.9985
$f =$	0.5000	0.1000	0.0100	
$i =$	140.0000	550.0000	1302.0000	
$u =$	156	856	4,040	
$v =$	1,333	6,667	66,667	
AFI =	0.5523	0.2024	0.0666	
Pa{p=.0015}	0.8953	0.8863	0.9429	

	$f = 1/2$ and $i = 140$			$f = 1/10$ and $i = 550$			$f = 1/100$ and $i = 1302$		
p	u	v	Pa	u	v	Pa	u	v	Pa
0.0010	150	2,000	0.9301	734	10,000	0.9316	2,679	100,000	0.9739
0.0015	156	1,333	0.8953	856	6,667	0.8863	4,040	66,667	0.9429
0.0020	162	1,000	0.8608	1,004	5,000	0.8328	6,276	50,000	0.8885
0.0025	168	800	0.8266	1,185	4,000	0.7715	10,010	40,000	0.7998
0.0030	174	667	0.7927	1,407	3,333	0.7032	16,331	33,333	0.6712
0.0035	181	571	0.7594	1,680	2,857	0.6298	27,161	28,571	0.5127
0.0040	188	500	0.7266	2,016	2,500	0.5536	45,912	25,000	0.3526
0.0045	196	444	0.6944	2,433	2,222	0.4774	78,675	22,222	0.2202
0.0050	203	400	0.6628	2,950	2,000	0.4040	136,377	20,000	0.1279
0.0060	220	333	0.6020	4,397	1,667	0.2749	421,313	16,667	0.0381
0.0070	239	286	0.5444	6,662	1,429	0.1766	1.34E+06	14,286	0.0106
0.0080	260	250	0.4904	10,238	1,250	0.1088	4.35E+06	12,500	0.0029
0.0090	283	222	0.4400	15,930	1,111	0.0652	1.44E+07	11,111	0.0008
0.0100	308	200	0.3934	25,056	1,000	0.0384	4.82E+07	10,000	0.0002
0.0150	486	133	0.2151	271,566	667	0.0024	2.34E+10	6,667	0.0000
0.0200	796	100	0.1116	3.35E+06	500	0.0001	1.33E+13	5,000	0.0000
0.0250	1,345	80	0.0561	4.46E+07	400	0.0000	8.28E+15	4,000	0.0000
0.0300	2,337	67	0.0277	6.29E+08	333	0.0000	5.57E+18	3,333	0.0000
0.0350	4,160	57	0.0135	9.25E+09	286	0.0000	3.99E+21	2,857	0.0000
0.0400	7,560	50	0.0066	1.41E+11	250	0.0000	3.03E+24	2,500	0.0000
0.0450	13,984	44	0.0032	2.21E+12	222	0.0000	2.41E+27	2,222	0.0000
0.0500	26,266	40	0.0015	3.57E+13	200	0.0000	2.02E+30	2,000	0.0000
0.0600	96,355	33	0.0003	1.00E+16	167	0.0000	1.62E+36	1,667	0.0000
0.0700	369,209	29	0.0001	3.09E+18	143	0.0000	1.55E+42	1,429	0.0000
0.0800	1.47E+06	25	0.0000	1.03E+21	125	0.0000	1.76E+48	1,250	0.0000
0.0900	6.03E+06	22	0.0000	3.74E+23	111	0.0000	2.37E+54	1,111	0.0000
0.1000	2.55E+07	20	0.0000	1.47E+26	100	0.0000	3.77E+60	1,000	0.0000

16.17.

Compare the plans developed in Exercise 16.16 in terms of average fraction inspected and their operating-characteristic curves. Which plan would you prefer if $p = 0.0375$?

Plan A: AFI = 0.5165 and $P_a\{p = 0.0375\} = 0.6043$
Plan B: AFI = 0.5272 and $P_a\{p = 0.0375\} = 0.4925$

Prefer Plan B over Plan A since it has a lower P_a at the unacceptable level of p.

Notes

Notes

Notes

Notes

Notes

Notes

Notes

Notes

Notes

Notes

Notes

Notes